NOW 2 kNOW™

Calculus I

by T. G. D'Alberto

Pithy Professor Publishing Company

Brighton, CO

Published by

Pithy Professor Publishing Company, LLC
PO Box 33824
Northglenn, CO 80233

ISBN: 978-0988205406

Library of Congress Control Number: 2012914946

Printed in the United States of America

About the Author

Dr. Tiffanie G. D'Alberto has a Ph.D. in Electrical & Computer Engineering from Cornell University and a B.S. and M.S. in Electrical Engineering from Virginia Polytechnic Institute & State University.

She has worked for over a decade in the telecommunications and aerospace industries as a scientist, program manager, and supervisor. She has engaged in numerous opportunities for tutoring, teaching, and mentoring throughout her career and schooling.

In her spare time, Tiffanie enjoys oil painting, drawing, reading, sewing, and running. She's a huge fan of Star Trek, Renaissance Festivals, and animals.

Tiffanie lives in Colorado with her fiancé, Colin, and their many wonderful pets.

Dedication

To my dearest Colin, who inspires me, encourages me, and supports me. I could never thank you enough.

To my high school Calculus teacher, Mr. Klima, who showed me how easy Calculus can be.

Acknowledgements

I always thank my family first: My parents for the foundation, the push, and the belief in me all along; My fiancé for his inspiration, encouragement, and unending support.

A huge thanks goes to my many excellent math teachers from middle school to high school to college that not only taught the material but also taught a new way of thinking necessary to excel in these subjects.

I'd also like to acknowledge **Calculus with Analytical Geometry**, Second Alternate Edition by Earl W. Swokowski. His clear, accurate, and thorough discussion of Calculus has aided in my understanding of the subject and in the development of this book.

Finally, I'd like to thank Amazon.com for their excellent publish-on-demand service that enables books such as these, and you, the reader, for making this investment in your future.

Table of Contents

Introduction

Welcome!

I've taken quite a number of math-intensive technical courses - most went very well, but a few went very poorly. Through many years of schooling, tutoring, and teaching, of successes as well as failures, I've learned what does and does not work in terms of learning math. This book is shaped after those lessons.

The philosophy of this book is three-fold:

1. **To excel at math is to understand math.** For example, you know how to play Go Fish. You not only know the rules, you understand the object of the game and the techniques that are required to dominate against your 4-year old opponent. It doesn't matter that this time you have a different hand. It doesn't matter that you haven't played in 10 years. You *understand* the game, so you can play it well. That's how you should learn math.
2. **To understand math, you need the story** . The story is the logic flow that allows you to keep building on your *understanding*. If someone tells you a story and skips a critical part of the plot, you would and should say, "Hey, back up!"
3. **To understand math, you also need the big picture.** The big picture is the outline of the logic, or story, placed in an area small enough for you to see it in its entirety. Like a file directory on a computer, it organizes the information. Once you see the flow of the big picture, it's easier for you to put the details of the story into their proper places.

I know there are many people out there who say, "I'm just not a math person." I've tutored a number of you, and it only took 1-2 sessions before you were off getting A's on your own. The key to learning math is not memorization, it's understanding. Be open to changing the way you think. Once you get the flow, you'll get the A's. I wish you great luck!

Layout:

The layout of this text is different from most academic books:

1. **The problem sets are saved to the end of the book.** While it is true that you should master one chapter before continuing to the next, and mastery takes practice, placing problem sets between chapters tends to take away from the story. In this book, you can read from beginning to end to understand the logical progression of the course, or stop to do problem sets as you desire.

2. **Solution sets give the critical steps to get the answers, not just the answers.** Because this is not a textbook for a classroom, there is no need to keep the "secret sauce" from you.

3. **Appendix A is an overall summary of the entire book.** It helps you visualize the big picture and logic flow to give you a framework into which you can organize the details.

In addition, the following visual markers will help you navigate the material...

Key terms defined for the first time are **bolded** and also found in the index.

Important equations are shown as:

> *important equations*

Illustrative graphics and additional notes are shown on the side to accompany the text.

Finally, examples are given as supplements to the text as well as for illustration:

> *Example* This is an example to illustrate a point or to give further definition. Skip it if you feel very comfortable with the material presented thus far.

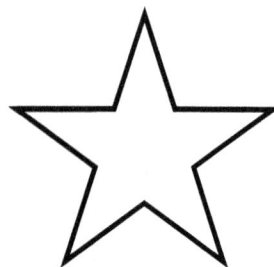

Notation

The following notation is often helpful when talking about Math. Some of this notation will be used in the text.

Closed interval – includes points a and b: [a,b]

Open interval – up to but not including points a and b: (a,b)

$\exists$	There exists
$\mathcal{R}$ or $\mathcal{R}$	With respect to
$\forall$	For all; For each; For every
$\ni$	Such that
$\in$	Is an element of
\	Except
iff or $\Leftrightarrow$	If and only if
b/c	Because
b/w	Between
f^n	Function
∞	On the order of (think 3 O's)
$\rightarrow$	Implies, Is given by
$\therefore$	Therefore
α	Proportional to
$\sim$ or $\approx$	Equivalent to; Approximately equal to
$\equiv$	Is assigned to; Is defined as; Is forced to equal to
$\neq$	Is not equal to
//	Parallel to
$\perp$	Perpendicular to; Orthogonal to

Final Note: variables are usually u, v, w, x, y, z, and constants are usually a, b, c, and k

Chapter 1: Lines & Functions

Lines:

Let's get started with a little review...

Lines can be defined by at least two points. The two points can be written as (x_1, y_1) and (x_2, y_2).

The **distance** between two points is given by:

$$distance = d = \sqrt{(x_2 - x_1)^2 + (y_2 - y_1)^2}$$

The **midpoint** between two points is given by:

$$midpoint = \left(\frac{x_1 + x_2}{2}, \frac{y_1 + y_2}{2}\right)$$

The **slope** between two points is given by:

$$slope = m = \frac{\Delta y}{\Delta x} = \frac{y_2 - y_1}{x_2 - x_1}$$

A line defined by two points:
(x_1, y_1) and (x_2, y_2).

Recall that the slope is the direction the line is heading if traveling from left to right. In the above equations, it does not matter which point you pick as (x_1, y_1) as long as you are consistent. Also, you can pick any two points on a straight line to calculate the slope.

Let's solve the slope equation for $\Delta y = y_2 - y_1$:

$$y_2 - y_1 = m(x_2 - x_1)$$

This is the **Point-Slope Form** for the equation of a line because it contains two points and a slope.

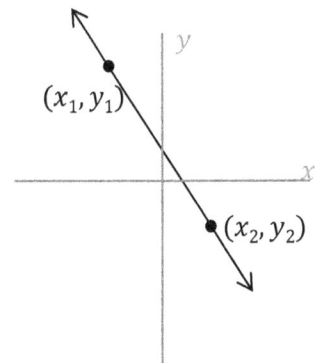

Now consider the special case of choosing (x_1, y_1) as the point where the line crosses the y-axis. This forces $x_1 = 0$, and we'll assign $y_1 \equiv b$. Now $(x_1, y_1) \equiv (0, b)$, and b is called the **y-intercept** of the line.

We can also choose any point on the line as (x_2, y_2), so to make that clear, we'll drop the subscripts and assign $(x_2, y_2) \equiv (x, y)$.

Plugging these two points into the Point-Slope Form of the equation for a line gives:

$$y_2 - y_1 = m(x_2 - x_1)$$

$$y - b = m(x - 0)$$

$$\boxed{y = mx + b}$$

This is the **Slope-Intercept Form** for the equation of a line as it contains the slope and the y-intercept.

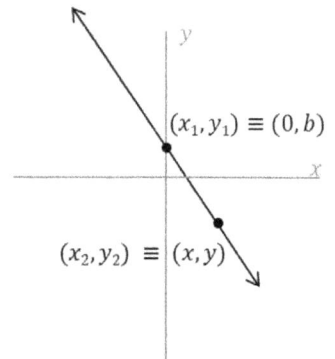

Defining one point on the line as the point where $x = 0$.

Example

For a **horizontal line**, all y's are equal, so $y_2 - y_1$ is always 0 which forces $m = 0$. The equation for the line becomes $y = b$.

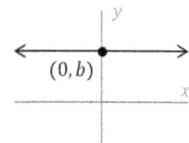

Horizontal line.

Example

For a **vertical line**, all x's are equal, so $x_2 - x_1$ is always 0 which forces $m = \infty$. There is no point-slope or slope-intercept form for the equation of a vertical line. The equation is simply $x = c$.

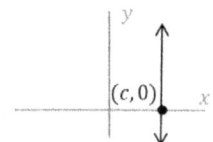

Vertical line.

Two lines are **parallel** iff $m_1 = m_2$.

Example

In other words, if the slope of line 1 is equal to the slope of line 2, then the two lines are parallel.

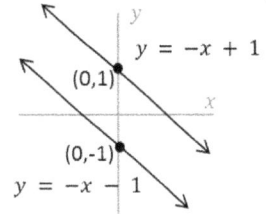

$y = -x + 1$
$(0,1)$
$(0,-1)$
$y = -x - 1$

Parallel lines.

Two lines are **perpendicular** iff $m_1 = -1/m_2$.

Example

In other words, if the slope of line 1 is the negative inverse of that of line 2, then the two lines are perpendicular.

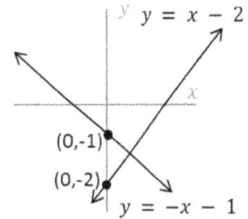

$y = x - 2$
$(0,-1)$
$(0,-2)$
$y = -x - 1$

Perpendicular lines.

Functions:

Definition: $f(x) = y$ is a **function** iff for every x value there exists only one y value. In other words, functions must pass the vertical line test – if you sweep a vertical line across the x-axis, the function should only intersect that line once.

The notation $y = f(x)$ means that y is a function of x.

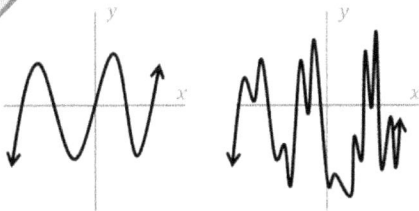

Example

These are functions.

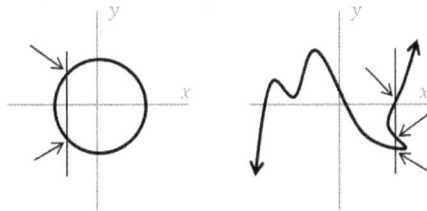

These are NOT functions.

A **polynomial** function is defined as a function that takes the following form:

$$y = a_n x^n + a_{n-1} x^{n-1} + \cdots + a_3 x^3 + a_2 x^2 + a_1 x + a_0$$

for all $a_i \in \{\mathbb{R}\}$ (for all real coefficients).

The **order** of the polynomial is the largest x-exponent, n, in the function. Common names are given to the first three orders:

> A **linear** function has order 1.
> A **quadratic** function has order 2.
> A **cubic** function has order 3.

Example

$y = x^5 + 3x^2$; order 5 polynomial

$y = 4x^2 + 2$; order 2 polynomial
   or quadratic function

$y = mx + b$; order 1 polynomial
   or linear function

A **composite function** is a function of a function written as:

$$(g \circ f)(x) = g(f(x))$$

Where $g(x)$ and $f(x)$ are each functions. In the above equation, you evaluate $f(x)$ and then plug $f(x)$ into the equation for $g(x)$ making it $g(f(x))$. Note that $(f \circ g)(x)$ and $(g \circ f)(x)$ aren't necessarily the same.

Example

If $g(x) = 4x^2$ and $f(x) = 2x + 5$ then:

$(g \circ f)(x) = 4 \cdot (2x + 5)^2 = 16x^2 + 80x + 100$

$(f \circ g)(x) = 8x^2 + 5$

Chapter 2: Limits & Tangents

Limits:

Like it sounds, the **limit** of a function is the boundary a function has at a given x value. The limit is written as:

$$limit = L = \lim_{x \to a} f(x)$$

where a is some value of x.

In English, this equation says literally that L is the limit or boundary of $f(x)$ as x approaches the value a.

You could also say that as x approaches the value a, $f(a)$ approaches some value L.

Example

$$y = f(x)$$
$$\lim_{x \to \infty} f(x) = 0$$
$$\lim_{x \to 0} f(x) = -\infty$$

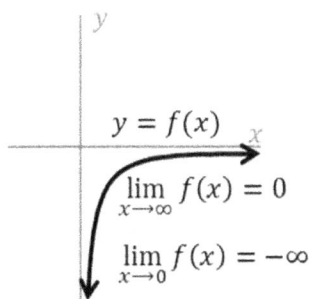

As the value of x approaches ∞, the function approaches $y = 0$;

As the value of x approaches 0, the function approaches $y = (-\infty)$.

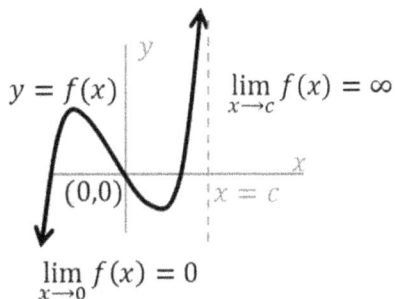

$$y = f(x)$$
$$\lim_{x \to c} f(x) = \infty$$
$$(0,0) \quad x = c$$
$$\lim_{x \to 0} f(x) = 0$$

As the value of x approaches c, the function approaches $y = \infty$;

As the value of x approaches 0, the function has a value of $y = 0$.

You can also talk about approaching **the limit from the left or the right**. The notation goes as follows:

$$L^- = \lim_{x \to a^-} f(x) \text{ from the left}$$

$$L^+ = \lim_{x \to a^+} f(x) \text{ from the right}$$

Note: If $\lim_{x \to a^-} f(x) = \lim_{x \to a^+} f(x) = k$,

then $\lim_{x \to a} f(x) = k$.

In other words, if you approach $x = a$ from the left giving $y = f(a) \to k$, and if you approach $x = a$ from the right giving the same answer of $y = f(a) \to k$, then it doesn't matter from which direction you approach – If you approach $x = a$, then $f(a)$ approaches the value k.

Example

$$y = \begin{cases} -1, x \in (-\infty, 1] \\ 2, x \in (1, \infty) \end{cases}$$

$$\lim_{x \to 1^-} f(x) = -1$$

$$\lim_{x \to 1^+} f(x) = 2$$

As the value of x approaches 1 from the left, $y = (-1)$;

As the value of x approaches 1 from the right, $y = 2$.

$$y = x^2$$

$$\lim_{x \to 0^-} f(x) = 0$$

$$\lim_{x \to 0^+} f(x) = 0$$

As the value of x approaches 0 from the left, $y = 0$;

As the value of x approaches 0, from the right, $y = 0$.

Tangents:

Recall that the slope of a line is given by:

$$slope = m = \frac{\Delta y}{\Delta x} = \frac{y - y_1}{x - x_1}$$

where (x, y) has been substituted for (x_2, y_2).

We can choose any (x_1, y_1) on the line $y = f(x)$ to fit in the slope equation, so let's choose $x_1 = a$ and find the corresponding y_1 by substituting into $y = f(x)$ giving us $y_1 = f(a)$.

The slope can then be written as:

$$slope = m = \frac{\Delta y}{\Delta x} = \frac{f(x) - f(a)}{x - a}$$

Because we can also choose any point (x, y) on the line, imagine choosing an x very close to a, written as $x \rightarrow a$, or x approaches a. (Obviously, if $x = a$, we only have one point, and one point alone does not define a line.)

We can then write the slope as:

$$slope = m = \lim_{x \to a} \frac{f(x) - f(a)}{x - a}$$

The above equation gives the **instantaneous slope** because it is the slope at the point $(a, f(a))$, where x is infinitesimally close to a.

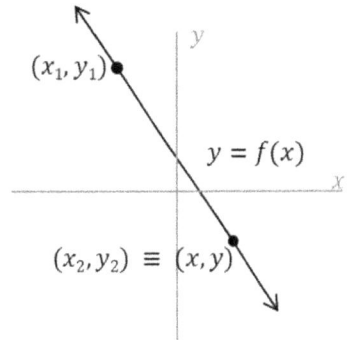

Using two points to find the slope of a line and making (x_2, y_2) generic.

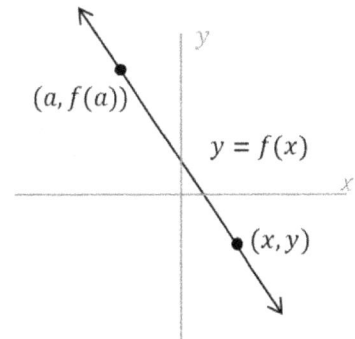

Setting $(x_1, y_1) \equiv (a, f(a))$.

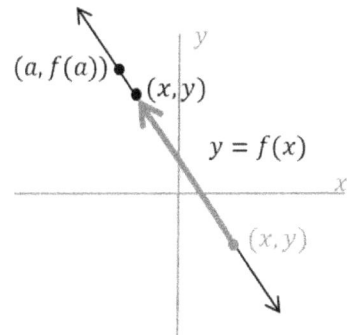

Moving (x, y) as close to $(a, f(a))$ as possible without overlapping to find instantaneous slope.

This may seem like a lot of effort when dealing with a straight line, but we can apply the same process to a polynomial function. Though a polynomial is curvy rather than straight, we can describe how those curves are changing at each point on the function.

Using the same process as before for straight lines, you can pick a point, $(a, f(a))$, on the curve described by $y = f(x)$. Next pick a point close to $(a, f(a))$ and draw a line.

As you pick your second point closer and closer to $(a, f(a))$, you'll notice that the line you draw eventually just skims the curve. This line is called the **tangent** to the curve.

The equation for the tangent of a curve at a given point $(a, f(a))$ can be written using any form for the equation of a line. For purposes to become clear in the next chapter, we will use the familiar slope intercept form:

$$y = mx + b$$

where m is the instantaneous slope.

Example

As point $(x, y) \rightarrow (a, f(a))$, the line between the two points becomes the tangent of the curve at point $(a, f(a))$.

You can also say that near the point $(a, f(a))$, the tangent line will intersect the curve at only one point.

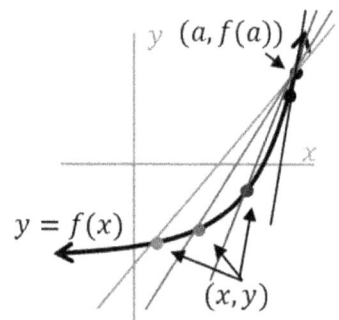

Finding the tangent at $(a, f(a))$.

Example ➤ Consider the slope of $y = x^2$ at the point where $x = 2$.

$$m = \lim_{x \to 2} \frac{f(x) - f(2)}{x - 2} = \lim_{x \to 2} \frac{x^2 - 4}{x - 2}$$

$$= \lim_{x \to 2} \frac{(x-2)(x+2)}{x-2} = \lim_{x \to 2} (x + 2) = 4$$

Note that in the above calculation, we didn't just substitute $x = 2$ right away. That would have given us 0/0 which is indeterminate.

To fit the y-intercept form of the tangent line equation, we have m and now need b – the y-intercept of the tangent line.

We can use the point-slope form of the equation for a line with $(x_1, y_1) = (2, 2^2) = (2,4)$, $m = 4$, and $(x_2, y_2) = (0, b)$. Solving for b:

$$y_2 - y_1 = m(x_2 - x_1)$$
$$(b - 4) = 4(0 - 2)$$
$$\to b = -12$$

So, the tangent line at $x = 2$ is given by the equation: $y = 4x - 12$.

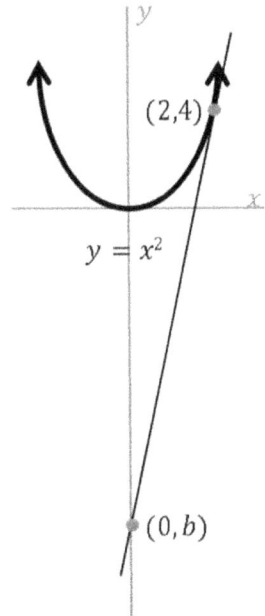

Tangent curve for $y = x^2$ at the point where $x = 2$.

You can imagine that if we can do this for one point on a function, we can do it for all points on a continuous function. This would become laborious indeed. What a nice segue to our next chapter.

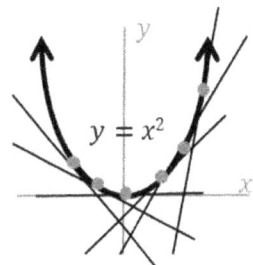

Tangent curves shown for various points on the $y = x^2$ curve.

13

Chapter 3: Derivatives

Derivatives:

Now we can finally start talking about calculus! The **derivative of $f(x)$ at a** is given by $f'(a)$:

$$f'(a) = \lim_{x \to a} \frac{f(x) - f(a)}{x - a}$$

This is just the instantaneous slope equation. Taking the derivative of something means finding its tangent or rate of change.

If the derivative $f'(a)$ exists, the function is said to be **continuous** at the point a. **Continuity** of a function means that the function can be drawn without having to lift your pencil.

As mentioned in the prior chapter, we want to find the tangent of every point on the curve $y = f(x)$. To make things more general, we're going to get rid of terms that focus on the particular point $(a, f(a))$.

Let's instead think of the more generic point, $(x, f(x))$, and a point very close to it called $(x + h, f(x + h))$. Here we're using h to represent a very small distance from x.

Using similar logic as was used to find the instantaneous slope equation, we want to find the ratio of $\Delta y / \Delta x$ between these two points as $h \to 0$. Putting it all together, we get the general equation for the **derivative of $f(x)$**:

$$f'(x) = \lim_{h \to 0} \frac{f(x+h) - f(x)}{h}$$

By using this form, it's easier to see that as we decrease h, we approach any point, (x, y), on the function $y = f(x)$.

One can arrive at the same equation by simply substituting $a = (x + h)$. The term $x \to a$ becomes $x \to x + h$ or $h \to 0$. The denominator, $x - a$, becomes $x - (x + h) = (-h)$, and the numerator is $f(x) - f(x + h)$.

$$f'(x) = \lim_{h \to 0} \frac{f(x) - f(x + h)}{-h}$$

$$= \lim_{h \to 0} \frac{f(x + h) - f(x)}{h}$$

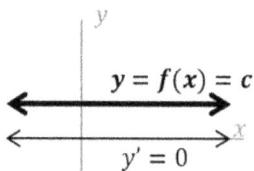

Consider the line $y = f(x) = c$;

$$f(x + h) = c;$$

$$f'(x) = \lim_{h \to 0} \frac{c - c}{h} = \lim_{h \to 0} \frac{0}{h} = 0$$

The slope of the function $y = c$ is 0 everywhere.

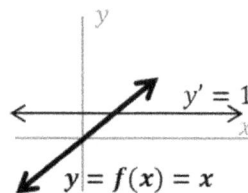

Consider the line $y = f(x) = x$;

$$f(x + h) = x + h$$

$$f'(x) = \lim_{h \to 0} \frac{x + h - x}{h} = \lim_{h \to 0} \frac{h}{h} = 1$$

The slope of the function $y = x$ is 1 everywhere.

Consider the function $y = f(x) = x^2$;

$$f(x + h) = (x + h)^2;$$

$$f'(x) = \lim_{h \to 0} \frac{(x+h)^2 - x^2}{h}$$

$$= \lim_{h \to 0} \frac{x^2 + 2hx + h^2 - x^2}{h} = \lim_{h \to 0} 2x + h = 2x$$

The slope of $y = x^2$ is $2x$: negative for $x < 0$, positive for $x > 0$, and always increasing. If you want to know the slope of $f(x)$ when $x = 2$, it is simply $f'(2) = 2x = 2 \times 2 = 4$. At $x = 0$, $f'(0) = 0$, and at $x = -2$, $f'(-2) = -4$.

Look at the pictures to the right to convince yourself that this method gives the instantaneous slope of the function everywhere.

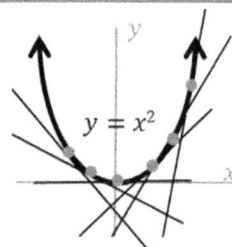

Individual tangent curves drawn for $y = x^2$.

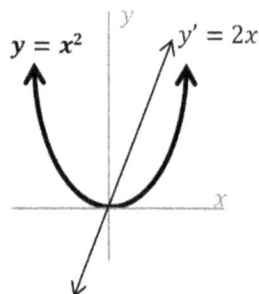

All of the tangent lines to $y = x^2$ have a slope of $f'(x) = 2x$.

This takes a lot work, but luckily, people have been doing this for a long time and patterns have emerged. The following pages summarize these patterns, or rules, which we can use as short-cuts.

Basic Rules for Derivatives:

Derivatives of Constants are 0
$$D_x[c] = 0$$

y=10	$y' = 0$
y=-14,256	$y' = 0$

Power Rule
$$D_x[x^n] = nx^{n-1}$$

y=x	$y' = 1 \cdot x^0 = 1$
y=x²	$y' = 2 \cdot x^1 = 2x$
y=x³	$y' = 3 \cdot x^2$
y=1/x = x⁻¹	$y' = -1 \cdot x^{-2} = -1/x^2$

Pull Out Constants
$$D_x[cf(x)] = cD_x[f(x)]$$

y=5x	$y' = 5(1 \cdot x^0) = 5$
y=10x²	$y' = 10(2 \cdot x^1) = 20x$
y=4x³	$y' = 4(3 \cdot x^2) = 12x^2$

Distribution
$$D_x[f(x) \pm g(x)] = D_x[f(x)] \pm D_x[g(x)]$$

y = 4x³ + 10x² - 5x

$y' = 4(3 \cdot x^2) + 10(2 \cdot x^1) - 5(1 \cdot x^0)$
$= 12x^2 + 20x - 5$

Product Rule
$$D_x[f(x) \cdot g(x)] = f(x)D_xg(x) + g(x)D_xf(x)$$
$$= fg' + gf'$$

y = 4x³ (10x² - 5x)

$y' = 4(3 \cdot x^2)(10x^2 - 5x)$
$\quad + 4x^3 (10(2 \cdot x^1) - 5(1 \cdot x^0))$
$= 12x^2 (10x^2 - 5x) + 4x^3(20x - 5)$
$= 120x^4 - 60x^3 + 80x^4 - 20x^3$
$= 200x^4 - 80x^3$

Quotient Rule
$$D_x\left[\frac{f(x)}{g(x)}\right] = \frac{g(x)D_xf(x) - f(x)D_xg(x)}{[g(x)]^2} = \frac{gf' - fg'}{g^2}$$

y = (10x² - 5x) / 4x³

$y' = \dfrac{4x^3(20x - 5) - (10x^2 - 5x)(12x^2)}{16x^6}$

$= \dfrac{-40x^4 + 40x^3}{16x^6}$

$= \dfrac{5(1 - x)}{2x^3}$

Note: It is easiest to remember the quotient rule as follows:

$$D_x\left[\frac{HI}{HO}\right] = \frac{HO \; dHI - HI \; dHO}{HO \; HO}$$

The derivative then reads:
  "HO dee HI minus HI dee HO, all over HO-HO."

HO's are on the outside in both the numerator and denominator.

Trigonometric Rules:

$$D_x[\sin(x)] = \cos(x) \qquad D_x[\cos(x)] = -\sin(x)$$
$$D_x[\tan(x)] = \sec^2(x) \qquad D_x[\sec(x)] = \sec(x)\tan(x)$$
$$D_x[\cot(x)] = -\csc^2(x) \qquad D_x[\csc(x)] = -\csc(x)\cot(x)$$

Example

Show $D_x[\tan(x)] = \sec^2(x)$:

$$D_x[\tan(x)] = D_x\left[\frac{\sin(x)}{\cos(x)}\right]$$

$$= \frac{\cos(x)(\cos(x)) - \sin(x)(-\sin(x))}{\cos^2(x)} \qquad \text{Quotient Rule}$$

$$= \frac{\cos^2(x) + \sin^2(x)}{\cos^2(x)} = \frac{1}{\cos^2(x)} = \sec^2(x) \qquad \text{Trig Properties}$$

Example

Show $D_x[\sin(x)] = \cos(x)$:

$$D_x[\sin(x)] = \lim_{h \to 0} \frac{\sin(x+h) - \sin(x)}{h}$$

Plot of $\sin(h)/h$

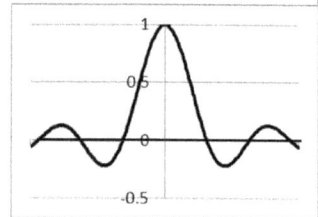

$$= \lim_{h \to 0} \frac{\sin(x)\cos(h) + \sin(h)\cos(x) - \sin(x)}{h}$$

$$= \lim_{h \to 0} \frac{\sin(x)(\cos(h) - 1) + \sin(h)\cos(x)}{h}$$

$$= \lim_{h \to 0} \frac{\sin(h)\cos(x)}{h}$$

$$= \cos(x) \cdot \lim_{h \to 0} \frac{\sin(h)}{h} \qquad \text{Special function called sync(h)} \atop \text{See sidebar for how to analyze}$$

$$= \cos(x)$$

$$\lim_{h \to 0^-} \frac{\sin(h)}{h} = 1;$$
$$\lim_{h \to 0^+} \frac{\sin(h)}{h} = 1;$$
$$\therefore \lim_{h \to 0} \frac{\sin(h)}{h} = 1$$

18

Chain Rule:

Let $y = f(u)$ and $u = g(x)$, i.e. $y = f(g(x))$.

$$\frac{dy}{dx} = \frac{dy}{du} \cdot \frac{du}{dx} = f'(u)du$$

In other words, take the derivative of the outside times the derivative of the inside.

Example

$y = (4x + 2)^2$

Let $u = 4x + 2$, then $du = 4$; so $y = u^2$
$y' = 2u \cdot du = 2(4x + 2) \cdot 4 = 32x + 16$ — Chain Rule

OR you could do it without the Chain Rule:
$y = 16x^2 + 16x + 4$ — Distribution
$y' = 32x + 16$

Example

$y = (2x + 1)^3$

Let u = 2x + 1, then du = 2; so y = u³
$y' = 3u^2 \cdot du = 3(2x + 1)^2 \cdot 2$ — Chain Rule
$= 6(2x + 1)^2$

OR you could do it without the Chain Rule:
$y = (2x + 1)^3 = 8x^3 + 12x^2 + 6x + 1$ — Distribution
$y' = 24x^2 + 24x + 6 = 6(4x^2 + 4x + 1)$
$= 6(2x + 1)^2$

Example

$y = \sin(2x)$

Let $u = 2x$, then $du = 2$; so $y = \sin(u)$ — Chain Rule
$y' = \cos(u) \cdot du = 2\cos(2x)$

OR you could do it without the Chain Rule:
$\sin(x + x) = \sin(x)\cos(x) + \cos(x)\sin(x)$ — Product Rule &
$= 2\sin(x)\cos(x)$ — Trig Properties

$y' = 2[\sin(x)(-\sin(x)) + \cos(x)(\cos(x))]$
$= 2[\cos^2(x) - \sin^2(x)] = 2\cos(2x)$

Chapter 4: Max & Min

Higher Order Derivatives:

Before we talk about how to find max's and min's, we need to look at **higher order derivatives**. For example, the second order derivative, $2°$ derivative or y'', is the derivative of the derivative of y. It is sometimes called the curvature of y.

> *Example*
>
> $y = 3x^2 + 5x + 2;$
>
> $y' = 6x + 5;$
>
> $y'' = 6$

Notation:

$1°$	$2°$	$3°$	$4°$	...	$n°$
y'	y''	y'''	$y^{(4)}$	...	$y^{(n)}$
$f'(x)$	$f''(x)$	$f'''(x)$	$f^{(4)}(x)$	...	$f^{(n)}(x)$
$D_x y$	$D_x^2 y$	$D_x^3 y$	$D_x^4 y$	...	$D_x^n y$
$\dfrac{dy}{dx}$	$\dfrac{d^2y}{dx^2}$	$\dfrac{d^3y}{dx^3}$	$\dfrac{d^4y}{dx^4}$	...	$\dfrac{d^ny}{dx^n}$

When dealing with functions of time, $x = f(t)$, the first derivative is called **velocity**, $v(t) = f'(t)$, and the second derivative is called **acceleration**, $a(t) = f''(t)$.

The primes and ()'s on the first two rows differentiate higher order *derivatives* from higher order *exponents*:

$y^{(4)} = f^{(4)}(x) = $ the 4th derivative of $y = f(x)$

$y^4 = f^4(x) = [f(x)]^4 = f(x) \cdot f(x) \cdot f(x) \cdot f(x)$

Local Extrema:

Consider a function $f(x)$ with $x_1 < x_2$.

Where $f(x_1) < f(x_2)$, $f(x)$ is increasing: **positive slope, $f'(x) > 0$.**

Where $f(x_1) > f(x_2)$, $f(x)$ is decreasing: **negative slope, $f'(x) < 0$.**

Where $f(x_1) = f(x_2)$, $f(x)$ is constant: **zero slope, $f'(x) = 0$.**

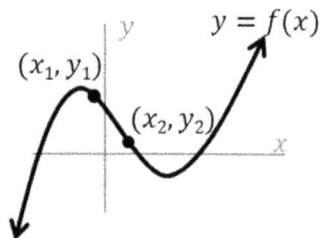

Two points on the curve are chosen such that $x_1 < x_2$. Since $y_1 > y_2$, $f(x)$ is decreasing between these two points.

A **local extreme** of $f(x)$ is when the function goes from increasing to decreasing or vice versa in a given range.

The local extreme is also referred to as a **critical point** or a **point of inflection**.

As a function goes through a point of inflection, its slope typically changes sign from positive to negative or vice versa. At the exact point of inflection, the slope is zero.

So, to find a point of inflection on $f(x)$, take the first derivative and find where it is equal to zero.

Example

Find the critical points for $y = x^2$;

$$y' = 2x$$

Set $y' = 2x = 0$ → Inflection occurs at $x = 0$.

To find the y-coordinate, plug $x = 0$ into $y = x^2$ → inflection occurs at $(0,0)$.

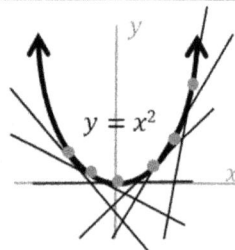

Inflection Point occurs at $x = 0$ when the tangent or slope = 0.

Find critical points: $y = 2x^3 + 3x^2 - 12x + 5$

Example

$$y' = 6x^2 + 6x - 12$$
$$= 6(x^2 + x - 2) = 6(x - 1)(x + 2)$$

Set $y' = 0 \rightarrow x = +1, -2$

Plug these x values into $f(x) \rightarrow$
inflections occur at $(1, -2)$ and $(-2, 25)$.

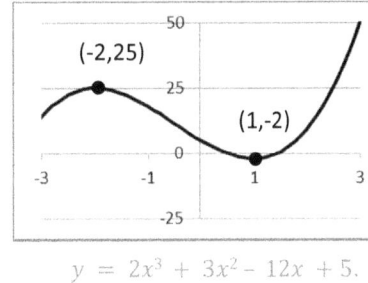

$y = 2x^3 + 3x^2 - 12x + 5.$

Max or Min:

Now we know where the inflection points are, but we have no way of knowing (without plotting) whether those points are **maxima** or **minima**. Here's where the second derivative becomes important.

Recall that the first derivative shows the slope of the original function. The second derivative shows the slope of the slope. It will let us know whether we have a minimum or maximum at the inflection point.

For illustration, consider $y = x^2$. We know $y' = 2x$ which means that there is a point of inflection at $(0,0)$. We also know that when $x < 0$ the slope (y') is negative, when $x > 0$ the slope is positive. And, we know that the slope is always increasing.

More simply, we can calculate $y'' = 2$ which tells us that the 2° derivative, or slope of the slope, is always positive.

If $f(x)$ is to go from decreasing to increasing, from negative slope to positive slope, then the slope of the slope must be increasing. This means that the second derivative must be greater than 0. When $y'' > 0$, we are at a local minimum.

Conversely, when $y'' < 0$, we are at a local maximum: the original function goes from positive slope to negative slope.

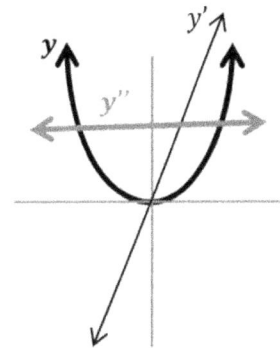

For the function $y = x^2$, setting the first derivative $y' = 2x \equiv 0$ tells us that the inflection occurs at $(0,0)$.

But, notice that the function $y' = 2x$ goes from negative to positive and is always increasing - the slope of y' is positive.

Indeed, $y'' = 2$ which is always greater than 0. Therefore, the point $(0,0)$ occurs at a local minimum.

23

Example

Find max & min: $y = 2x^3 + 3x^2 - 12x + 5$

$y' = 6x^2 + 6x - 12$

We determined that inflections occurred at $(1, -2)$ and $(-2, 25)$.

Now take the 2° derivative:
$y'' = 12x + 6$

Plug in $x = 1 \rightarrow 12 \cdot 1 + 6 = 18 > 0 \rightarrow$ min
Plug in $x = -2 \rightarrow 12 \cdot -2 + 6 = -18 < 0 \rightarrow$ max

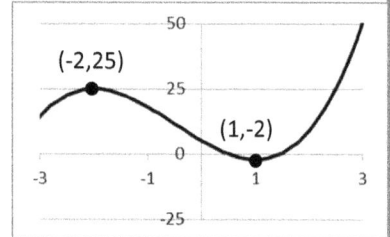

(-2,25)

(1,-2)

$y = 2x^3 + 3x^2 - 12x + 5.$

Example

Find max & min: $y = \sin(2x)$

$y' = 2\cos(2x)$

$y' = 0$ when $2x = \pm\pi/2$, or $x = \pm\pi/4 \sim 0.79$
  Plug into original equation:
  Inflection occurs at $(\pi/4, 1)$ and $(-\pi/4, -1)$
  (For completeness, inflection occurs at every
  $x = n(\pi/4)$ where n is any odd integer.)

$y'' = -4\sin(2x)$

At $x = \pi/4, y'' = -4 < 0 \rightarrow$ max
At $x = -\pi/4, y'' = +4 > 0 \rightarrow$ min

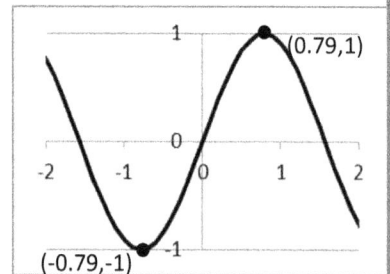

(0.79,1)

(-0.79,-1)

$y = \sin(2x).$

In summary:

To find an inflection point:
  Set $y' = 0$ and solve for x.

To determine max or min:
  Evaluate y'' at your x values.
    If $y''(x_i) > 0 \rightarrow$ minimum
    If $y''(x_i) < 0 \rightarrow$ maximum

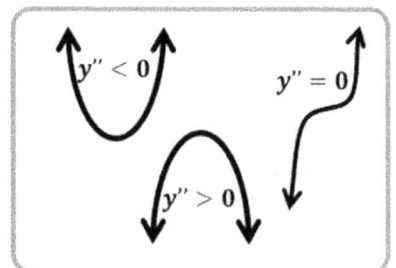

$y'' < 0$ $y'' = 0$

$y'' > 0$

Second Derivative Test.

Chapter 5: Anti-Derivatives

Anti-Derivatives:

Anti-derivatives undo derivatives. To perform an anti-derivative, think – if I can go forward, I can go back.

If $2x$ is the derivative of x^2, then x^2 is the anti-derivative of $2x$.

But, so is $x^2 + 5, x^2 + 100$, and $x^2 - \pi$ (since $D_x[c] = 0$).

Therefore, to be general, we say that the anti-derivative of $2x$ is $x^2 + c$, where c is some arbitrary constant.

Notation:

The anti-derivative in symbolic form is $F(x)$. As shown in the above example, the anti-derivative is given by:

$$F(x) = f(x) + c$$

Find the anti-derivative: $f(x) = 5x^3$;

Intuitively, $F(x) = \frac{5x^4}{4} + c$

Check:

$$D_x\left(F(x)\right) = D_x\left(\frac{5x^4}{4} + c\right) = 4\left(\frac{5x^3}{4}\right) + 0 = 5x^3$$

The next page will give us rules for how to actually find the anti-derivative so you don't have to just guess.

Basic Rules for Anti-Derivatives:

Power Rule

If $y = x^n$; $\qquad\qquad F(x) = \frac{x^{n+1}}{n+1} + c$

Pull Out Constants

If $y = kf(x)$; $\qquad\qquad F(x) = k \times F\left[f(x)\right] + c$

Distribution

If $y = f(x) \pm g(x)$; $\qquad F(x) = F(x) \pm G(x) + c$

$\qquad$ where $G(x)$ is the anti-derivative of $g(x)$

Compound Functions
$\qquad$ Let $u = g(x)$ such that $u' = g'(x) = k$ (a constant), then

$$F(x) = \frac{F(u)}{u'} + c$$

Example

Find the anti-derivative: $y = 10$.

$$F(x) = 10\left[\frac{x^{0+1}}{0+1}\right] + c = 10x + c$$

Check: $D_x[10x + c] = 10 + 0 = 10$

Example

Find the anti-derivative: $y = x^2 - x$.

$$F(x) = \frac{x^{2+1}}{2+1} - \frac{x^{1+1}}{1+1} + c = \frac{x^3}{3} - \frac{x^2}{2} + c$$

Check: $D_x\left[\frac{x^3}{3} - \frac{x^2}{2} + c\right] = \frac{3x^2}{3} - \frac{2x}{2} + 0 = x^2 - x$

Find the anti-derivative: $y = 5(2x + 1)^4$.

Example

Let $u = (2x + 1)$; then $u' = 2$ which is a constant. Now we can use the formula for compound functions:

$F(x) = \frac{5F[u]}{u'} + c$ and $F(u) = \frac{u^5}{5}$ so

$F(x) = \frac{5(2x+1)^5}{5}\left(\frac{1}{2}\right) + c = \frac{(2x+1)^5}{2} + c$

Check: $D_x\left[\frac{(2x+1)^5}{2} + c\right] = D_x\left[\frac{(u)^5}{2}\right] + 0$

$= \frac{1}{2}(5u^4)du = \frac{1}{2}(5(2x + 1)^4) \times 2 = 5(2x + 1)^4$

Anti-Derivatives of Trigonometric Functions:

If $y = \sin(x)$; $F(x) = -\cos(x) + c$
If $y = \cos(x)$; $F(x) = \sin(x) + c$

Find the anti-derivative: $y = \cos(5x)$.

Example

Let $u = 5x$; $u' = 5$ which is a constant, and $F(u) = \sin(u) + c$;

$F(x) = \frac{F(u)}{u'} + c = \frac{\sin(u)}{5} + c = \frac{\sin(5x)}{5} + c$

Check: $D_x\left[\frac{\sin(5x)}{5} + c\right] = D_x\left[\frac{\sin(u)}{5}\right] + 0$

$= \frac{1}{5}(\cos(u)\,du) = \frac{1}{5}\cos(5x) \times 5 = \cos(5x)$

Chapter 6: Integrals

Integrals:

What you learned in Chapter 5 about taking anti-derivatives is also called taking the **integral**. The integral has the notation:

$$F[x] = \int f(x)dx$$

There are two types of integrals: the indefinite integral and the definite integral.

The **indefinite integral** looks exactly like taking the general anti-derivative with the familiar unknown constant added to the end:

$$\int f(x)dx = F(x) + c$$

The **definite integral** is taken when we have additional information to figure out what c really is:

$$\int_a^b f(x)dx = F(x)\Big|_a^b = F(b) - F(a)$$

Note that with the definite integral we have the additional information of a and b. These values are called **the lower bound** (a) and **upper bound** (b) of the integral.

In practice, to evaluate the definite integral, you take the anti-derivative, $F(x)$, as normal then evaluate it at $x = a$ and at $x = b$. The difference $F(b) - F(a)$ is the answer to the definite integral.

$f(x) = 5x^2 + 4, x \in [0,4]$
$\quad$ (x goes from $x = 0$ to $x = 4$)

$F(x) = \int_0^4 (5x^2 + 4)dx = \left(\frac{5x^3}{3} + 4x\right)\big|_0^4$

$= \left(\frac{5(4)^3}{3} + 4(4)\right) - \left(\frac{5(0)^3}{3} + 4(0)\right)$

$= \frac{5}{3}(64) + 16 - 0 = 122.7$

$f(x) = x^3 + 3, x \in [-2,2]$

$F(x) = \int_{-2}^2 (x^3 + 3)dx = \left(\frac{x^4}{4} + 3x\right)\big|_{-2}^2$

$= \left(\frac{(2)^3}{4} + 3(2)\right) - \left(\frac{(-2)^3}{4} + 3(-2)\right)$

$= \frac{8}{4} + 6 - \frac{-8}{4} - (-6) = 12$

$f(x) = 2\sin(5x), x \in [0, \pi/2]$

$F(x) = \int_0^{\pi/2} (2\sin(5x))dx = 2\left(\frac{-\cos(5x)}{5}\right)\big|_0^{\pi/2}$

$= 2\left(\frac{-\cos(\frac{5\pi}{2})}{5} - \frac{-\cos(0)}{5}\right) = 2\left(\frac{-0}{5} - \frac{-1}{5}\right) = \frac{2}{5}$

The Significance of Integrals:

Yippideeyaya. We can turn a function into a number with a definite integral. What's all the fuss about?

Well, the definite integral is really giving you the area between a curve and the x-axis. The integral is a summation of infinitesimally small rectangles the height of which follow the line of $f(x)$. One could write the integral as:

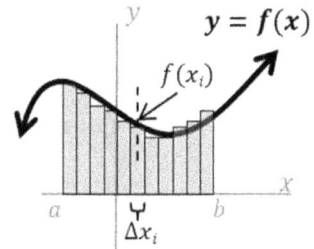

$$\int_a^b f(x)dx = \lim_{n \to \infty} \sum_{i=1}^n f(x_i)\Delta x_i; \quad x\epsilon[a,b]$$

The term Δx_i means delta x, or the difference between x_i and x_{i-1}. This Δx_i is the width of one of the rectangles, and $f(x_i)$ is the height. Summing the areas of each rectangle, $f(x_i)\cdot\Delta x_i$, gives you the total area under the curve. And, as you divide up the integral $[a,b]$ into more and more rectangles ($n \to \infty$), you get a better and better approximation of the area.

The area between a curve and the x-axis can be approximated by summing the areas of a series of rectangles on the integral $[a,b]$ that have a width Δx_i and a height $f(xi)$.

Note that as Δx_i gets smaller, it is replaced by the symbol dx, the subscripts can be dropped, and the integrand symbol replaces the limit and sum.

Example ▸ Consider a basic triangle as shown. We know the area is given by
$$A = \frac{1}{2}bh,$$
where b is the base, and h is the height. In this case, we can see that
$$A = \frac{1}{2}(3)(3) = \frac{9}{2}$$

Now let's do it with calculus:
$$y = x; \quad x \in [0,3]$$

$$F(x) = \int_0^3 xdx = \frac{x^2}{2}\Big|_0^3 = \frac{3^2}{2} - \frac{0^2}{2} = \frac{9}{2}$$

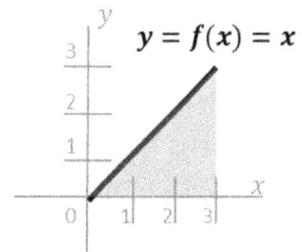

Triangle bounded by $y = x$ and the x-axis from $x = 0$ to $x = 3$.

Find the area under the curve $y = x^2 + 2$; $x \in [-1,2]$

You could take a bunch of rectangles and sum them up, but let's just use calculus:

$$A = F(x) = \int_{-1}^{2}(x^2 + 2)dx = \left(\frac{x^3}{3} + 2x\right)\Big|_{-1}^{2}$$

$$= \frac{2^3}{3} + 2(2) - \frac{(-1)^3}{3} - 2(-1) = \frac{8}{3} + 4 + \frac{1}{3} + 2 = 9$$

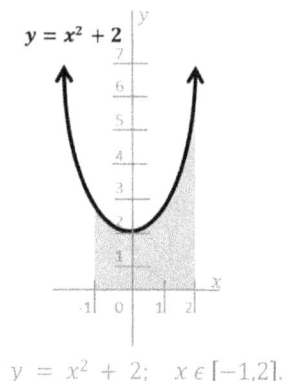

$y = x^2 + 2$

$y = x^2 + 2; \quad x \in [-1,2]$.

Find the area between the curves:

$y = x + 2$ and $y = x^2$; $\quad x \in [0,2]$

At first this may seem hard, but if you know how to break the problem into parts, it becomes quite simple.

We know that we can find the area between each function and the x-axis. So, if we take the area of the highest function minus the area of the lowest function, we should be left with the area between the two functions.

$$A_{Total} = A_1 - A_2$$

$$A_1 = F_1(x) = \int_0^2 (x + 2)dx = \left[\frac{x^2}{2} + 2x\right]\Big|_0^2$$
$$= \frac{2^2}{2} + 2(2) - \frac{0^2}{2} - 2(0) = 6$$

$$A_2 = F_2(x) = \int_0^2 (x^2)dx = \frac{x^3}{3}\Big|_0^2$$
$$= \frac{2^3}{3} - \frac{0^3}{3} = {}^8/_3$$

$$A_{Total} = 6 - {}^8/_3 = {}^{10}/_3$$

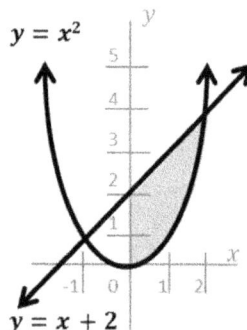

$y = x^2$

$y = x + 2$

Area between two functions.

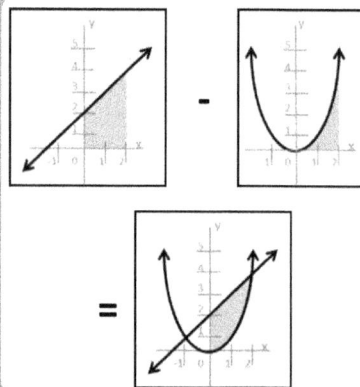

Breaking the problem into parts.

Chapter 7: More on Integrals

Properties of Integrals:

Change Sign (and Flip Boundaries)

$$\int_a^b f(x)dx = -\int_b^a f(x)dx$$

Pull Out Constants

$$\int_a^b cf(x)dx = c\int_a^b f(x)dx$$

Distribution

$$\int_a^b [f(x) \pm g(x)]dx = \int_a^b f(x)dx \pm \int_a^b g(x)dx$$

Break Intervals
For $a < k < b$;

$$\int_a^b f(x)dx = \int_a^k f(x)dx + \int_k^b f(x)dx$$

The above rules are very helpful in analyzing integrals. The last rule in particular is many times key to avoiding wrong answers. The following examples will illustrate how useful these properties can be. One thing will become clear: graphing is a must when evaluating definite integrals!

Example

Find the area of $y = \begin{cases} x & [0,2) \\ 2 & [2,4] \end{cases}$

$$Area = \int_0^2 xdx + \int_2^4 2dx = \frac{x^2}{2}\Big|_0^2 + 2x\Big|_2^4$$

$$= [2 - 0] + [8 - 4] = \underline{6}$$

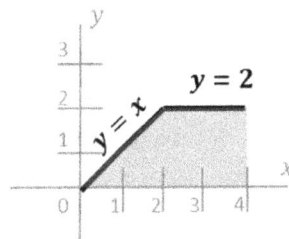

To analyze this area, break the interval into two parts: one for the triangle, one for the square.

Find the area of
$$y = \sin(x); \; x \in [-\pi/2, \pi/2]$$

Wrong way:
$$\int_{-\pi/2}^{\pi/2} \sin(x)dx = -\cos(x)\Big|_{-\pi/2}^{\pi/2} = -[0-0] = 0$$

Right way:
Here is a subtlety that can often be overlooked when not graphing functions prior to evaluating them in definite integrals. Recall that the integral finds the area between a function and the x-axis. This is the same as writing:
$$\int_a^b [f(x) - g(x)]dx \text{ where } g(x) = 0.$$

However, if the x-axis is above the function, then the correct integral to use is:
$$\int_a^b [g(x) - f(x)]dx \text{ where } g(x) = 0.$$

So, for our example we note that $y = 0$ is above $y = \sin(x)$ for $x < 0$, and the reverse is true for $x > 0$. As a result, we break the interval into two parts:

$$\int_{-\pi/2}^{\pi/2} \sin(x)dx$$

$$= \int_{-\pi/2}^{0} [0 - \sin(x)]dx + \int_{0}^{\pi/2} [\sin(x) - 0]dx$$

$$= +\cos(x)\Big|_{-\pi/2}^{0} + -\cos(x)\Big|_{0}^{\pi/2}$$

$$= [1 - 0] - [0 - 1] = \underline{2}$$

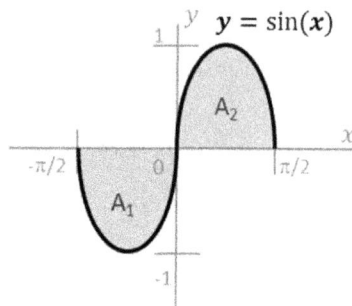

Finding the area of $y = \sin(x)$ between $-\pi/2$ and $\pi/2$ requires breaking the interval into two parts.

Example

Find the area between the lines:

$y = x - 1$; $y = -2x - 4$ on $x \in [-3,2]$

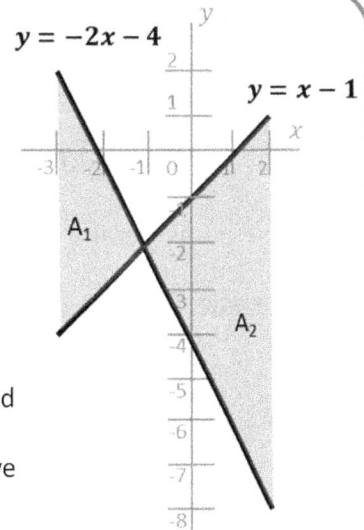

$y = -2x - 4$

$y = x - 1$

Wrong way:

$$A = \int_{-3}^{2}[x - 1 - (-2x - 4)]dx$$
$$= \int_{-3}^{2}[3x + 3]dx = \frac{3x^2}{2} + 3x \Big|_{-3}^{2}$$
$$= [6 + 6] - [13.5 - 9] = 7.5$$

Right way:

There is actually more than one right way. We could do this the same way we did the last example of Chapter 6. We can find the area of $y = x - 1$ above and below the x-axis and do the same for $y = -2x - 4$. We can then break up the integrals to add and subtract the areas correctly. Unfortunately, this requires eight interval evaluations!

The short way to evaluate the problem.

Let's do this more directly. Note that these lines cross at $x = -1$; Also note that the function $y = x - 1$ is on the bottom for $x < -1$ and on the top for $x > -1$. With that information, we can proceed with just two integrals:

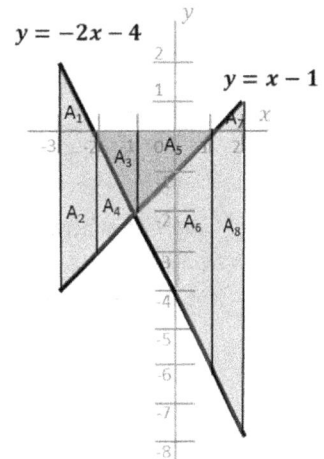

$y = -2x - 4$

$y = x - 1$

$$A_1 = \int_{-3}^{-1}[-2x - 4 - (x - 1)]dx$$
$$= \int_{-3}^{-1}[-3x - 3]dx = \frac{-3x^2}{2} - 3x \Big|_{-3}^{-1}$$
$$= [-1.5 + 3] - [-13.5 + 9] = 6$$

$$A_2 = \int_{-1}^{2}[x - 1 - (-2x - 4)]dx$$
$$= \int_{-1}^{2}[3x + 3]dx = \frac{3x^2}{2} + 3x \Big|_{-1}^{2}$$
$$= [6 + 6] - [1.5 - 3] = 13.5$$

$A = A_1 + A_2 - A_3 + A_4 - A_5 + A_6 + A_7 + A_8$

The long way to evaluate the problem.

$$A_{total} = A_1 + A_2 = 6 + 13.5 = \underline{19.5}$$

Two Useful Techniques:

Functions of Functions

$$F(u) = \int f(u)du$$

where $u = g(x), du = g'(x)$

Integration by Parts

$$\int u\,dv = uv - \int v\,du$$

where $u = g(x), du = g'(x)$
and $v = h(x), dv = h'(x)$

These two techniques help reduce difficult integrals into simple problems. The key is to know how to look for the patterns. Practice will help you find these patterns quickly.

For the following examples, we'll consider indefinite integrals so that we can see the symbolic form of the answer more easily.

Example

Evaluate:
$$\int x(x^2 + 1)^3 dx$$

We know how to find the integral of $\int u^3 du$, so...
let $u = x^2 + 1$, then $du = 2xdx$. (Putting the "dx" at the end is stating the Chain Rule from Chapter 3.)

Looking at the integrand, all we're missing is the 2 to use the Functions of Functions technique. Now we can write:

$$\int x(x^2 + 1)^3 dx = \frac{1}{2}\int 2x(x^2 + 1)^3 dx$$

$$= \frac{1}{2}\int u^3 du = \frac{1}{2}\left[\frac{u^4}{4}\right] + c = \frac{1}{8}[x^2 + 1]^4 + c$$

Note: Though we can't put x's in the integrand willy-nilly to make life convenient, we can certainly use constants (see the Pull Out Constants property earlier in this chapter).

Check: $D_x\left[\frac{1}{8}[x^2 + 1]^4 + c\right]$
$$= \frac{4}{8}[x^2 + 1]^3(2x) = x[x^2 + 1]^3$$

Example

Evaluate:
$$\int \sin(5x)\,dx$$

Let $u = 5x$, then $du = 5dx$.

Again, we're only off from the original problem by a constant, so we can write:

$$\int \sin(5x)\,dx = \frac{1}{5}\int 5\sin(5x)\,dx$$
$$= \frac{1}{5}\int \sin(u)\,du = \frac{-1}{5}\cos(u) + c$$

$$= \frac{-1}{5}\cos(5x) + c$$

Check: $D_x\left[\frac{-1}{5}\cos(5x) + c\right]$
$$= \frac{-1}{5}[-\sin(5x)](5) = \sin(5x)$$

Example

Evaluate:
$$\int x\sin(5x)\,dx$$

For this problem, we'll have to try integration by parts: $\int u\,dv = uv - \int v\,du$.

Now we have an extra x, and we can't use the Functions of Functions technique to pull x's or functions of x across the integral barrier.

From practice, we learn that we want to make the $\int v\,du$ term as simple as possible. So, lets make the following assignments:
$$u = x;$$
$$dv = \sin(5x)\,dx$$

Therefore :
$$du = dx;$$
$$v = \int \sin(5x)\,dx = \frac{-1}{5}\cos(5x)$$

Here we'll drop the constant of integration, c, but we'll be sure to add it at the end.

$$\int u\,dv = uv - \int v\,du$$
$$= x \cdot \frac{-1}{5}\cos(5x) - \int \frac{-1}{5}\cos(5x)\,dx$$

$$= \frac{-1}{5}x\cos(5x) + \frac{1}{5}\left(\frac{1}{5}\right)\int 5\cos(5x)\,dx$$

$$= \frac{-1}{5}x\cos(5x) + \frac{1}{25}\sin(5x) + c$$

Check:
$$D_x\left[\frac{-1}{5}x\cos(5x) + \frac{1}{25}\sin(5x) + c\right]$$
$$= \frac{-1}{5}[-5x\sin(5x) + \cos(5x)]$$
$$+ \frac{5}{25}\cos(5x)$$
$$= x\sin(5x)$$

Example

Evaluate:
$$\int x^2 \sin(5x)\, dx$$

For this problem, we'll have to try integration by parts again: $\int u\, dv = uv - \int v\, du$.

Let's make the following assignments:
$$u = x^2;$$
$$dv = \sin(5x)\, dx$$

Therefore :
$$du = 2x\, dx;$$
$$v = \int \sin(5x)\, dx = \frac{-1}{5}\cos(5x)$$

$$\int u\, dv = uv - \int v\, du$$
$$= x^2 \cdot \frac{-1}{5}\cos(5x) - \int \frac{-1}{5}(2x)\cos(5x)\, dx$$

For the second term, do it again:
$$u = x;$$
$$dv = \cos(5x)\, dx$$

$$du = dx;$$
$$v = \int \cos(5x)\, dx = \frac{1}{5}\sin(5x)$$

$$\frac{2}{5}\int x\cos(5x)\, dx$$

$$= \frac{2}{5}\left[x \cdot \frac{1}{5}\sin(5x) - \int \frac{1}{5}\sin(5x)\, dx\right]$$

$$= \frac{2}{5}\left[x \cdot \frac{1}{5}\sin(5x) + \frac{1}{25}\cos(5x)\right]$$

Put it all together:
$$\int x^2 \sin(5x)\, dx$$
$$= \frac{-x^2}{5}\cos(5x) + \frac{2}{5}\left[\frac{x}{5}\sin(5x) + \frac{1}{25}\cos(5x)\right] + c$$

Check:
$$D_x\left[\frac{-1}{5}x^2\cos(5x)\right]$$
$$= \frac{-1}{5}[2x\cos(5x) - 5x^2\sin(5x)]$$
$$= \frac{-2x}{5}\cos(5x) + x^2\sin(5x)$$

$$D_x\left[\frac{2}{25}x\sin(5x)\right]$$
$$= \frac{2}{25}[\sin(5x) + 5x\cos(5x)]$$
$$= \frac{2}{25}\sin(5x) + \frac{2x}{5}\cos(5x)$$

$$D_x\left[\frac{2}{125}\cos(5x) + c\right]$$
$$= \frac{-2}{25}\sin(5x)$$

Add them together and you get :

$$x^2\sin(5x).$$

Chapter 8: Logarithms & Exponentials

General Logarithms & Exponentials:

The **logarithm**, or log for short, of a number is given by:

$$\log_b y = x$$

where b is a constant called the base of the log. When $b = 10$, it is usually not written explicitly.

The inverse of a log is the **exponential** given by:

$$y = b^x$$

One way to remember this is to picture b getting on the log, sailing across the river, and hoisting x on its shoulders.

$$\log_b y = x$$

$$y \underset{b}{=\!=\!=} x$$

$$y =\!=\!= b^x$$

One way to remember how to perform a log function is to picture the logarithm as a log and the equal sign as a river.

Example

$\log 100 = x$	$\log_2 8 = x$
$100 = 10^x$	$8 = 2^x$
$x = 2$	$x = 3$

Because the exponential is the inverse of the log and vice versa, we say that they are **inverse functions** of each other. The notation for an inverse function is f^{-1}.

$$f^{-1}(\log_b x) = b^x; \qquad f^{-1}(b^x) = \log_b x$$

If you perform an inverse function on a function, you will get the argument as the answer:

$$b^{(\log_b x)} = x; \qquad \log_b(b^x) = x$$

$\log 10^2 = x$	$\log_2 2^3 = x$
$10^2 = 10^x$	$2^3 = 2^x$
$x = 2$	$x = 3$

We could rewrite the preceding examples this way to emphasize the inverse function property.

Natural Logarithms & Exponentials:

The special case of the logarithm where $b = e \approx 2.71828$ is called the **natural logarithm** and is given the symbol, **ln**.

$$\log_e x = \ln x$$

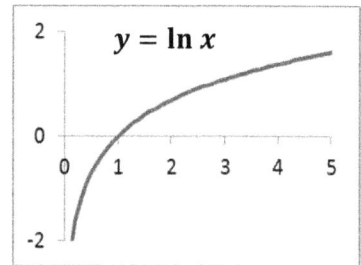

$y = \ln x$

Natural logarithm.

The inverse of the natural log is the **natural exponential** given by e^x or **exp(x)**:

$$f^{-1}(\ln x) = e^x = \exp(x); \quad f^{-1}(e^x) = \ln x$$

By definition of an inverse function we have:

$$e^{(\ln x)} = x; \quad \ln(e^x) = x$$
$$x > 0; \quad \text{for all } x$$

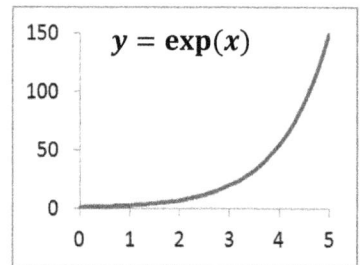

$y = \exp(x)$

Natural exponential.

Note that the argument of the natural log must be greater than 0. This is because $\ln(x)$ is only defined where $x > 0$.

This all may seem rather arbitrary to consider a special case where the base is equal to some weird value and call it natural. However, the natural log has a formal definition:

$$\ln a = \int_1^a \frac{1}{x} dx$$

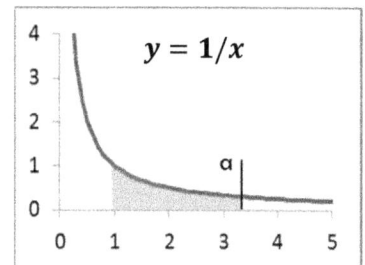

$y = 1/x$

Area of $1/x, x \in [1, a]$.

If you've noticed that you can't perform an anti-derivative on x^{-1} because it leads to a divide by zero error, you were not alone. Mathematicians noticed this, too, and came up with the natural log as a way to get around that problem. Only when $b = e$ does this relation hold.

Properties of Logarithms & Exponentials:

Logs and exponents have special properties that make
them rather user friendly. For example, multiplication and
division get replaced with addition and subtraction. The
following tables list the laws of logs and exponents. The
insets show the special case of the natural logs and natural
exponents.

$$\log_b(p)^r = r\log_b p$$

$$\log_b(pq) = \log_b p + \log_b q$$

$$\log_b\left(\frac{p}{q}\right) = \log_b p - \log_b q$$

$$\ln(p)^r = r\ln p$$

$$\ln(pq) = \ln p + \ln q$$

$$\ln\left(\frac{p}{q}\right) = \ln p - \ln q$$

$p, q > 0$
r is real

$$(a^p)^r = a^{pr}$$

$$a^p a^q = a^{p+q}$$

$$\frac{a^p}{a^q} = a^{p-q}$$

$$\left(\frac{a}{b}\right)^p = \frac{a^p}{b^p}$$

$$(e^p)^r = e^{pr}$$

$$e^p e^q = e^{p+q}$$

$$\frac{e^p}{e^q} = e^{p-q}$$

$$(ab)^p = a^p b^p$$

$a, b > 0$
p, q, r are real

Example		
$x = \log_2 64$	$x = \log_2 64$	
$= \log_2 8 + \log_2 8$	$= 2\log_2 8$	
$x = 3 + 3 = 6$	$x = 2 \cdot 3 = 6$	
$x = 2^6$	$x = 2^6$	
$= 2^3 \cdot 2^3$	$= (2^3)^2$	
$x = 8 \cdot 8 = 64$	$x = 8^2 = 64$	

Calculus with Logarithms & Exponentials:

By definition of the natural logarithm, we can deduce the derivative of ln. We can also use the Functions of Functions technique and the Chain Rule to generalize the result for $u = f(x) \neq 0$:

$$\int \frac{1}{x} dx = \ln|x| + c \qquad \int \frac{1}{u} du = \ln|u| + c$$

$$D_x[\ln|x|] = \frac{1}{x} \qquad D_x[\ln|u|] = \frac{1}{u} D_x[u]$$

In addition, there is a theorem that gives the derivative of the natural exponential function. Similar to above, we can generalize the result:

$$\int e^x dx = e^x + c \qquad \int e^u du = e^u + c$$

$$D_x[e^x] = e^x \qquad D_x[e^u] = e^u D_x[u]$$

Side Note:
You might recall that

$$\sin x = \frac{e^{ix} - e^{-ix}}{2i}; \text{ and}$$

$$\cos x = \frac{e^{ix} + e^{-ix}}{2};$$

where $i = \sqrt{-1}$.

Using the definition of the derivative of e^x:

$$D_x[\sin x] = D_x \left[\frac{e^{ix} - e^{-ix}}{2i} \right]$$

$$= \frac{1}{2i} \left[ie^{ix} - (-i)e^{-ix} \right]$$

$$= \frac{(e^{ix} + e^{-ix})}{2} = \cos x$$

Example

Find the derivative of $y = e^{(x^2+1)^2}$

Let $u = (x^2 + 1)^2$;
Then $du = 2(x^2 + 1)(2x) = 4x^3 + 4x$;

$$D_x \left[e^{(x^2+1)^2} \right] = (4x^3 + 4x)e^{(x^2+1)^2}$$

Example

Evaluate $\int \frac{x}{x^2+4} dx$

Let $u = x^2 + 4$; $du = 2x$;

$$\int \frac{x}{x^2+4} dx = \frac{1}{2} \int \frac{2x}{x^2+4} dx$$

$$= \ln|x^2 + 4| + c$$

To deal with general logs and exponents, we can use the definition of the inverse function and the properties of logs to come up with an alternate expression for a^x:

$$a^x = e^{\ln a^x} = e^{x \ln a}; \quad a > 0$$

We can analyze the derivative of a^x:

$$D_x[a^x] = D_x\left[e^{x \ln a}\right] = e^{x \ln a} D_x[x \ln a] = a^x \ln a$$

$$\underset{\text{Definition } a^x}{\quad} \quad \underset{D_x e^u = e^u du}{\quad} \quad \underset{D_x kx = k}{\quad}$$

Similarly, we can analyze the following derivative:

$$D_x\left[\frac{a^x}{\ln a}\right] = \frac{1}{\ln a} D_x[a^x] = \frac{1}{\ln a} a^x \ln a = a^x$$

$$\underset{D_x k f(x) = k D_x f(x) \quad \text{Above result}}{\quad}$$

Putting it all together and generalizing the results, we have:

$$\int a^x dx = \frac{a^x}{\ln a} + c \qquad \int a^u du = \frac{a^u}{\ln a} + c$$

$$D_x[a^x] = a^x \ln a \qquad D_x[a^u] = a^u \ln a\, D_x(u)$$

Finally, let's let $y = \log_a x$ which means $x = a^y$. Now take the ln of both sides of $x = a^y$ and differentiate:

$$\ln x = \ln a^y = y \ln a \;\rightarrow\; y = \frac{\ln x}{\ln a} = \log_a x$$

$$D_x y = D_x \log_a x = D_x \frac{\ln x}{\ln a} = \frac{1}{\ln a}\left[\frac{1}{x}\right]$$

$$D_x[\log_a x] = \frac{1}{x \ln a} \qquad D_x[\log_a u] = \frac{1}{u \ln a} D_x(u)$$

Evaluate $\int x 4^{-x^2} dx$

Example

Let $u = -x^2;\ du = -2x;$

$$\int x 4^{-x^2} dx = \frac{-1}{2} \int -2x\, 4^{-x^2} dx$$

$$= \frac{-1}{2} \int a^u du = \frac{-1}{2\ln 4} 4^{-x^2} + c$$

Evaluate $\int \ln x\, dx$

Example

Use Integration by Parts: $\int u\, dv = uv - \int v\, du$

$$u = \ln x; \qquad dv = dx;$$
$$du = \frac{1}{x} dx; \qquad v = x;$$

$$\int \ln x\, dx = x \ln x - \int \frac{x}{x} dx$$

$$\boxed{\int \ln x\, dx = x[\ln(x) - 1] + c}$$

Evaluate $\int \log_a x\, dx$

Example

Use Integration by Parts: $\int u\, dv = uv - \int v\, du$

$$u = \log_a x; \qquad dv = dx;$$
$$du = \frac{1}{x \ln a} dx; \qquad v = x;$$

$$\int \log_a x\, dx = x \log_a x - \int \frac{x}{x} \ln a\, dx$$

$$= x[\log_a x - \ln a] + c$$

$$\boxed{\int \log_a x\, dx = x\left[\frac{\ln x}{\ln a} - \ln a\right] + c}$$

Chapter 9: Common Trig Functions

Common Trig Functions:

Recall from Chapter 3 the derivatives of the common trigonometric functions:

$$D_x[\sin(u)] = \cos(u)du \qquad\qquad D_x[\cos(u)] = -\sin(u)du$$

$$D_x[\tan(u)] = \sec^2(u)du \qquad\qquad D_x[\sec(u)] = \sec(u)\tan(u)du$$

$$D_x[\cot(u)] = -\csc^2(u)du \qquad\qquad D_x[\csc(u)] = -\csc(u)\,\cot(u)du$$

From these rules, we can easily derive the following integrals:

$$\int \cos(u)\,du = \sin(u) + c \qquad\qquad \int \sin(u)\,du = -\cos(u) + c$$

$$\int \sec^2(u)\,du = \tan(u) + c \qquad\qquad \int \sec(u)\tan(u)du = \sec(u) + c$$

$$\int \csc^2(u)\,du = -\cot(u) + c \qquad\qquad \int \csc(u)\cot(u)du = -\csc(u) + c$$

With the help of natural logs, we can find more integrals:

$$\int \tan x\,dx = \int \frac{\sin x}{\cos x}dx = \int \frac{1}{u}du = -\ln|\cos x| = \ln\left|\frac{1}{\cos x}\right| \qquad u = \cos x;\ \ du = -\sin x\,dx$$

$$= \ln|\sec x| + c;$$

$$\int \sec x\,dx = \int \sec x\,\frac{\sec x + \tan x}{\sec x + \tan x}dx = \int \frac{\sec^2 x + \sec x\tan x}{\sec x + \tan x}dx \qquad \begin{array}{l} u = \sec x + \tan x; \\ du = \sec x\tan x + \sec^2 x \end{array}$$

$$= \int \frac{1}{u}du = \ln|\sec x + \tan x| + c;$$

This substitution probably took some mathematician years to figure out!

Using similar techniques, we can generate the following table:

$$\int \tan(u)\,du = \ln|\sec(u)| + c \qquad\qquad \int \cot(u)\,du = \ln|\sin(u)| + c$$

$$\int \sec(u)\,du \qquad\qquad\qquad \int \csc(u)du$$
$$= \ln|\sec(u) + \tan(u)| + c \qquad = \ln|\csc(u) - \cot(u)| + c$$

Inverse Common Trig Functions:

Just as the logarithms and exponentials are inverse functions of each other, there are also inverse trigonometric functions defined as follows:

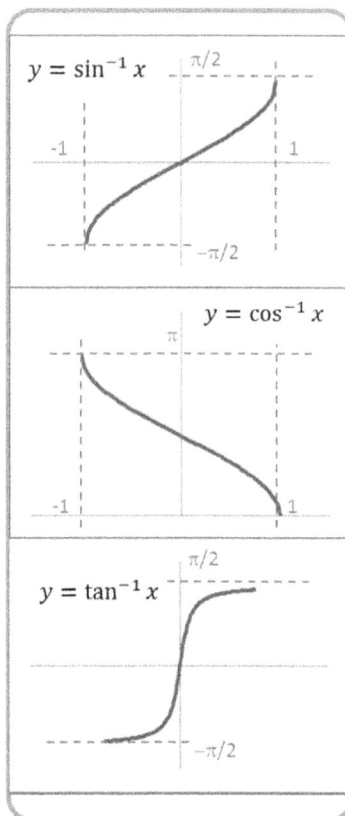

$$\sin^{-1}(\sin(x)) = x \qquad \cos^{-1}(\cos(x)) = x \qquad \tan^{-1}(\tan(x)) = x$$

$$\csc^{-1}(\csc(x)) = x \qquad \sec^{-1}(\sec(x)) = x \qquad \cot^{-1}(\cot(x)) = x$$

Note that the inverse trig functions can also be written with the exponent removed and the term "arc" used as a prefix instead. For example, $\sin^{-1}(x) = \arcsin(x)$. Also note that where the original function equals infinity, the inverse function is undefined.

So, let's find the derivatives of the inverse trig functions. We'll use $y = \sin^{-1}(x)$ as an example and leave the rest for practice problems.

Let $y = \sin^{-1}(x)$ which means $x = \sin(y)$ or
$$x = \sin(\sin^{-1}(x)).$$

Taking the derivative of both sides we get:

$$D_x[x] = D_x[\sin(\sin^{-1}(x))] = D_x[\sin(u)] = \cos(u)\, D_x[u]$$

$$1 = \cos(u)\, D_x[u];$$

From Trig, we know that $\sin^2(x) + \cos^2(x) = 1$, so

$$1 = D_x[u]\sqrt{1 - \sin^2(u)} = D_x[u]\sqrt{1 - \sin[\sin^{-1}(x)]^2}$$

$$= D_x[u]\sqrt{1 - x^2} = D_x[\sin^{-1}(x)]\sqrt{1 - x^2}$$

Solving for the derivative of the $\arcsin(x)$ we get:

$$D_x[\sin^{-1}(x)] = \frac{1}{\sqrt{1-x^2}}.$$

$y = \sin^{-1} x$

$y = \cos^{-1} x$

$y = \tan^{-1} x$

Plots of some of the inverse trig functions.

Using similar logic and generalizing with the Chain Rule:

$$D_x[\sin^{-1} u] = \frac{D_x[u]}{\sqrt{1-u^2}} = -D_x[\cos^{-1} u]$$

$$D_x[\tan^{-1} u] = \frac{D_x[u]}{1+u^2} = -D_x[\cot^{-1} u]$$

$$D_x[\sec^{-1} u] = \frac{D_x[u]}{u\sqrt{u^2-1}} = -D_x[\csc^{-1} u]$$

To integrate the inverse trig functions, we use these results along with integration by parts:

To evaluate $\int \sin^{-1}(x)\, dx$, use $\int u\,dv = uv - \int v\,du$:

$$u = \sin^{-1}(x) \qquad\qquad dv = dx;$$
$$du = \left.1\middle/\sqrt{1-x^2}\right.\, dx \qquad\qquad v = x$$

$$\int \sin^{-1}(x)\, dx = x\sin^{-1}(x) - \int \frac{x}{\sqrt{1-x^2}}\, dx$$

$$= x\sin^{-1}(x) + \frac{1}{2}\int \frac{2x}{\sqrt{1-x^2}}\, dx = x\sin^{-1}(x) + \frac{1}{2}\frac{\sqrt{1-x^2}}{1/2}$$

$$= x\sin^{-1}(x) + \sqrt{1-x^2} + c$$

Similarly we can get results for all but $\sec^{-1} x$ and $\csc^{-1} x$. Those results are included here for convenience and will be derived at the end of the next chapter.

$$\int \sin^{-1} u\, du = u\sin^{-1} u + \sqrt{1-u^2} + c$$

$$\int \cos^{-1} u\, du = u\cos^{-1} u - \sqrt{1-u^2} + c$$

$$\int \tan^{-1} u\, du = u\tan^{-1} u - \frac{1}{2}\ln\left|1+u^2\right| + c$$

$$\int \csc^{-1} u\, du = u\csc^{-1} u \pm \ln(u + \sqrt{u^2-1}) + c \quad \begin{cases} + \to 0 < \csc^{-1} u < \pi/2 \\ - \to -\pi/2 < \csc^{-1} u < 0 \end{cases}$$

$$\int \sec^{-1} u\, du = u\sec^{-1} u \mp \ln(u + \sqrt{u^2-1}) + c \quad \begin{cases} - \to 0 < \sec^{-1} u < \pi/2 \\ + \to \pi/2 < \sec^{-1} u < \pi \end{cases}$$

$$\int \cot^{-1} u\, du = u\cot^{-1} u + \frac{1}{2}\ln\left|1+u^2\right| + c$$

Example

Find $D_x\left[\sin^{-1}\left[\frac{x^2}{x+1}\right]\right]$.

$$D_x[\sin^{-1} u] = \frac{D_x[u]}{\sqrt{1-u^2}};$$

$$u = \frac{x^2}{x+1}; \quad du = \frac{(x+1)2x-x^2}{(x+1)^2} = \frac{x^2+2x}{x^2+2x+1};$$

$$D_x\left[\sin^{-1}\left[\frac{x^2}{x+1}\right]\right] = \frac{x^2+2x}{x^2+2x+1}\left[\frac{1}{\sqrt{1-\left(x^2/(x+1)\right)^2}}\right]$$

Example

Find $\int x \sin^{-1}(4x^2)\, dx$.

Let $u = 4x^2$; $du = 8x$;

$$\int x \sin^{-1}(4x^2)\, dx = \frac{1}{8}\int 8x \sin^{-1}(4x^2)\, dx$$

$$= \frac{1}{8}\left[4x^2 \sin^{-1}(4x^2) + \sqrt{1 - 16x^4}\right]$$

Example

Find $\int \frac{2}{1+4x^2}\, dx$.

Let $u = 2x$; $du = 2$;

$$\int \frac{2}{1+4x^2}\, dx = \int \frac{du}{1+u^2};$$

We know that $D_x[\tan^{-1} u] = \frac{D_x[u]}{1+u^2}$, so

$$\int \frac{2}{1+4x^2}\, dx = \tan^{-1}(2x) + c$$

Congratulations! This is an advanced technique covered in Chapter 11.

Chapter 10: Hyperbolic Trig Functions

Hyperbolic Trig Functions:

Now for something you might not have seen before – the **hyperbolic trigonometric functions**. By definition, the hyperbolic trig functions are:

$$\sinh x = \frac{e^x - e^{-x}}{2} \qquad \operatorname{csch} x = \frac{2}{e^x - e^{-x}}$$

$$\cosh x = \frac{e^x + e^{-x}}{2} \qquad \operatorname{sech} x = \frac{2}{e^x + e^{-x}}$$

$$\tanh x = \frac{e^x - e^{-x}}{e^x + e^{-x}} \qquad \coth x = \frac{e^x + e^{-x}}{e^x - e^{-x}}$$

Notice that once you know $\sinh x$ and $\cosh x$ (typically pronounced "sinch" and "cosh"), you can figure out the other functions.

Just as in trigonometry, you can derive relations between the hyperbolic functions. For example, let's take the square of $\sinh x$ and $\cosh x$:

$$\sinh^2(x) = \frac{1}{4}[e^{2x} - 2e^x e^{-x} + e^{-2x}]$$
$$= \frac{1}{4}[e^{2x} + e^{-2x} - 2]$$

$$\cosh^2(x) = \frac{1}{4}[e^{2x} + e^{-2x} + 2]$$

Now subtract $\sinh^2(x)$ from $\cosh^2(x)$ to get:

$$\cosh^2(x) - \sinh^2(x)$$
$$= \frac{1}{4}[e^{2x} + e^{-2x} + 2 - e^{2x} - e^{-2x} + 2] = 1$$

We can do the similar operations to derive the following relations between the hyperbolic trig functions:

$$\cosh^2(x) - \sinh^2(x) = 1$$

$$1 - \tanh^2(x) = \operatorname{sech}^2(x)$$

$$\coth^2(x) - 1 = \operatorname{csch}^2(x)$$

Note on nomenclature

The "common" trig functions as I've referred to them are actually called the **circular trig functions**. This is because of the relation:
$$\cos^2 x + \sin^2 x = 1.$$
If you use this equation to plot $\cos x$ versus $\sin x$, you get an equation that looks like $x^2 + y^2 = 1$ which is the plot of a circle.

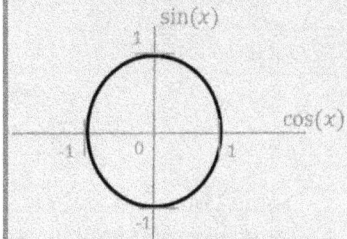

$$\cos^2 x + \sin^2 x = 1$$

Now if you consider the relation $\cosh^2 x - \sinh^2 x = 1$, and you plot $\cosh x$ versus $\sinh x$, you get something that looks like $x^2 - y^2 = 1$ which is the equation for a hyperbola.

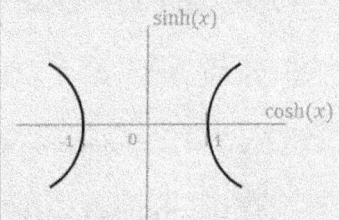

$$\cosh^2 x - \sinh^2 x = 1$$

This is where the term hyperbolic comes from to describe these functions.

49

Let's find the derivatives and integrals. The derivatives are very straightforward thanks to the ease of working with exponentials. Two derivations are given below:

$$D_x[\sinh x] = D_x\left[\frac{e^x - e^{-x}}{2}\right] = \frac{e^x + e^{-x}}{2} = \cosh x$$

$$D_x[\tanh x] = D_x\left[\frac{e^x - e^{-x}}{e^x + e^{-x}}\right] = D_x\left[\left(\frac{e^x - e^{-x}}{e^x + e^{-x}}\right)\left(\frac{e^x}{e^x}\right)\right]$$

$$= D_x\left[\left(\frac{e^{2x} - 1}{e^{2x} + 1}\right)\right] = \frac{(e^{2x}+1)(2e^{2x}) - (e^{2x}-1)(2e^{2x})}{(e^{2x}+1)^2}$$

$$= \frac{(4e^{2x})}{e^{4x} + 2e^{2x} + 1} = \frac{(4e^{2x})}{e^{4x} + 2e^{2x} + 1}\left(\frac{e^{-2x}}{e^{-2x}}\right) = \frac{4}{e^{2x} + 2 + e^{-2x}}$$

$$= \frac{2^2}{(e^x + e^{-x})^2} = sech^2 x$$

Some of the integration techniques are straightforward, some are not. For convenience, the following table summarizes the calculus properties of the hyperbolic trig functions:

$D_x(\sinh u) = \cosh u\ du$	$\int \sinh u\ du = \cosh u + c$
$D_x(\cosh u) = \sinh u\ du$	$\int \cosh u\ du = \sinh u + c$
$D_x(\tanh u) = sech^2 u\ du$	$\int \tanh u\ du = \ln|\cosh u| + c$
$D_x(\operatorname{csch} u) = -\operatorname{csch} u \coth u\ du$	$\int \operatorname{csch} u\ du = \ln\left|\tanh\left(\frac{u}{2}\right)\right| + c$
$D_x(\operatorname{sech} u) = -\operatorname{sech} u \tanh u\ du$	$\int \operatorname{sech} u\ du = 2\tan^{-1}(e^u) = \tan^{-1}(\sinh u) + c$
$D_x(\coth u) = -csch^2 u\ du$	$\int \coth u\ du = \ln|\sinh u| + c$

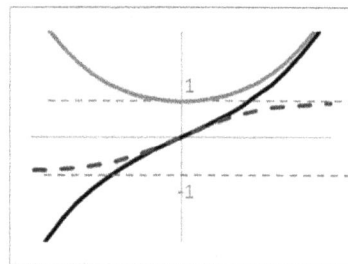

Plot of Hyperbolic functions:
——— sinh (x)
═══ cosh(x)
- - - tanh(x)

Example

Show $\int \tanh x \, dx = \ln|\cosh x| + c$.

$$\int \tanh x \, dx = \int \frac{\sinh x}{\cosh x} dx = \int \frac{du}{u} = \ln(u) + c \qquad u = \cosh x \; ; \quad du = \sinh x$$

$$= \ln|\cosh x| + c$$

Example

Show $\int \operatorname{sech} x \, dx = 2\tan^{-1}(e^x) + c$.

$$\int \operatorname{sech} x \, dx = \int \frac{2}{e^x + e^{-x}} dx = 2 \int \frac{1}{e^x + e^{-x}} \left(\frac{e^x}{e^x}\right) dx$$

$$= 2 \int \frac{e^x}{e^{2x}+1} dx = 2 \int \frac{du}{u^2+1} = 2\tan^{-1} u + c \qquad u = e^x \; ; \quad du = e^x$$

$$= 2\tan^{-1}(e^x) + c$$

Example

Evaluate $\int x \sinh(x^2) \cosh(x^2) \, dx$.

Let $u = \sinh(x^2)$; $du = 2x \cosh(x^2)$;

$$\frac{1}{2}\int 2x \sinh(x^2) \cosh(x^2) \, dx = \frac{1}{2}\int u \, du = \frac{u^2}{2} + c$$

$$= \frac{1}{4}\sinh^2 x + c$$

Note that one could also obtain $\frac{1}{4}\cosh^2 x + c$. Since we have the relation:

$$\cosh^2(x) - \sinh^2(x) = 1$$

we know that $\cosh^2(x)$ and $\sinh^2(x)$ only differ by a constant, 1. Therefore, the two answers should be equivalent because of the constant of integration, c. In this case, the c's should differ by 1/4 x 1 = 1/4.

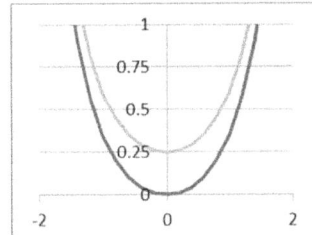

Plotted is $y = \sinh^2 x / 4$ in black and $y = \cosh^2 x / 4$ in grey. Note that they differ by a constant value of $^1/_4$.

Inverse Hyperbolic Trig Functions:

Indeed, as you would expect, there are functions which undo the hyperbolic trig functions called the **inverse hyperbolic trig functions**. They are defined in the usual way with $\sinh^{-1}(\sinh x) = x$, etc. A plot of some of those functions is shown to the right.

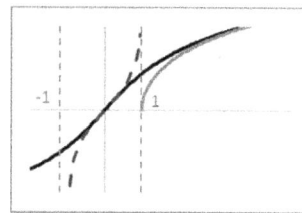

Plot of Hyperbolic functions:

———— $\sinh^{-1}(x)$
———— $\cosh^{-1}(x)$
- - - - $\tanh^{-1}(x)$

The inverse functions can be written in terms of natural logs as follows:

$$\sinh^{-1}x = \ln\left(x + \sqrt{x^2 + 1}\right)$$

$$\operatorname{csch}^{-1}x = \ln\left[\frac{1+\sqrt{1+x^2}}{x}\right]; \ x \in (0,1]$$

$$\cosh^{-1}x = \ln(x + \sqrt{x^2 - 1}); \ x \geq 1$$

$$\operatorname{sech}^{-1}x = \ln\left[\frac{1+\sqrt{1-x^2}}{x}\right]; \ x \in (0,1]$$

$$\tanh^{-1}x = \frac{1}{2}\ln\left[\frac{1+x}{1-x}\right]; \ x \in (-1,1)$$

$$\coth^{-1}x = \frac{1}{2}\ln\left[\frac{x+1}{x-1}\right]; \ x \in (-1,1)$$

The derivations of the above formulas are accomplished by using the exponential form of the hyperbolic function and solving for the inverse. As an example:

Example

Let $y = \sinh^{-1}x$. Then

$$x = \sinh y = \frac{e^y - e^{-y}}{2} = \frac{e^y - e^{-y}}{2}\left(\frac{e^y}{e^y}\right) = \frac{e^{2y} - 1}{2e^y}$$

$$e^{2y} - 2xe^y - 1 = 0$$

$$e^y = \frac{2x \pm \sqrt{4x^2 - 4(1)(-1)}}{2} = \frac{x + \sqrt{x^2 + 1}}{1}$$

$$y = \sinh^{-1}x = \ln\left|x + \sqrt{x^2 + 1}\right|$$

Use Pythagorean's Theorem:

$$x = \frac{-b \pm \sqrt{b^2 - 4ac}}{2a}$$

and drop the "-" before the radical to ensure $e^y > 0$

Using the symbolic forms of the inverse functions, we can, with a little bookkeeping, find the derivatives:

$$D_x[\sinh^{-1} u] = \frac{1}{\sqrt{u^2+1}} du \qquad\qquad D_x[\text{csch}^{-1}u] = \frac{-1}{u\sqrt{1+u^2}} du$$

$$D_x[\cosh^{-1}u] = \frac{1}{\sqrt{u^2-1}} du \qquad\qquad D_x[\text{sech}^{-1}u] = \frac{-1}{u\sqrt{1-u^2}} du$$

$$D_x[\tanh^{-1}u] = \frac{1}{1-u^2} du \qquad\qquad D_x[\coth^{-1}u] = \frac{1}{1-u^2} du$$

Example

Show $D_x[\tanh^{-1}x] = \frac{1}{1-x^2}$.

$$D_x[\tanh^{-1}x] = D_x\left[\frac{1}{2}\ln\left[\frac{1+x}{1-x}\right]\right]$$

$$= \frac{1}{2}\left[\frac{1-x}{1+x}\right]\left[\frac{(1-x)(1)-(1+x)(-1)}{(1-x)^2}\right] \qquad\qquad D_x(\ln u) = \frac{du}{u}$$

$$= \frac{1}{2}\left[\frac{1}{1+x}\right]\left[\frac{2}{1-x}\right] = \frac{1}{1-x^2}$$

Finally, we can turn to the integrals. For some we need integration by parts, for others we need to go back to the logarithmic form. A summary of the integrals is given as follows:

$$\int \sinh^{-1} u \; du = u\sinh^{-1}u - \sqrt{u^2+1} + c$$

$$\int \cosh^{-1}u \; du = u\cosh^{-1}u \mp \sqrt{u^2-1} + c \qquad \begin{cases} \longrightarrow \cosh^{-1}u > 0 \\ + \rightarrow \cosh^{-1}u < 0 \end{cases}$$

$$\int \tanh^{-1}u \; du = u\tanh^{-1}u + \frac{1}{2}\ln(1-u^2) + c$$

$$\int \text{csch}^{-1}u \; du = u\,\text{csch}^{-1}u + \sinh^{-1}u + c$$

$$\int \text{sech}^{-1}u \; du = u\,\text{sech}^{-1}u + \sin^{-1}u + c$$

$$\int \coth^{-1}u \; du = u\coth^{-1}u + \frac{1}{2}\ln(u^2-1) + c$$

Example

Show $\int \sinh^{-1} x\, dx = x\sinh^{-1} x - \sqrt{x^2+1} + c.$

Use integration by parts with:

$u = \sinh^{-1} x;$ $\qquad\qquad dv = dx;$

$du = \dfrac{1}{\sqrt{x^2+1}}\,dx;$ $\qquad v = x;$

$\int \sinh^{-1} x\, dx = x\sinh^{-1} x - \int \dfrac{x\,dx}{\sqrt{x^2+1}}$

$= x\sinh^{-1} x - \dfrac{1}{2}\int \dfrac{2x\,dx}{\sqrt{x^2+1}}$

$= x\sinh^{-1} x - \sqrt{x^2+1} + c$

Example

Show
$$\int \tanh^{-1}x\, dx = x\tanh^{-1}x + \frac{1}{2}\ln(1-x^2) + c.$$

For this derivation, we'll use the symbolic form:

$\int \tanh^{-1}x\, dx = \int \frac{1}{2}\ln\frac{1+x}{1-x}\,dx$

$= \frac{1}{2}\left[\int \ln(1+x)dx - \int \ln(1-x)dx\right]$

$= \frac{1}{2}[(1+x)(\ln(1+x)-1)$
$+ (1-x)(\ln(1-x)-1)] + c$ $\qquad \int \ln u\, du = u(\ln u - 1) + c$

$= \frac{1}{2}[x(\ln(1+x) - \ln(1-x))$
$+ (\ln(1+x) + \ln(1-x)] + c$

$= \frac{1}{2}\left[x\ln\left[\frac{1+x}{1-x}\right] + \ln[(1+x)(1-x)]\right] + c$

$= x\tanh^{-1}x + \frac{1}{2}\ln(1-x^2) + c$ $\qquad \tanh^{-1}x = \frac{1}{2}\ln\frac{1+x}{1-x}$

Example

Find $\int \sec^{-1} x \; dx$.

Use integration by parts with:

$u = \sec^{-1} x;$ $\qquad\qquad\qquad dv = dx;$

$du = \dfrac{1}{x\sqrt{x^2-1}} dx;$ $\qquad\quad v = x;$

$\int \sec^{-1} x \; dx = x \sec^{-1} x \; - \int \dfrac{dx}{\sqrt{x^2-1}}$

Recalling that $D_x[\cosh^{-1} u] = \dfrac{1}{\sqrt{u^2-1}} du$, we can now say:

$= x \sec^{-1} x \; - \cosh^{-1} x \; + c$

Or equivalently,

$= x \sec^{-1} x - \ln(x + \sqrt{x^2 - 1}) + c$

Example

Find $\int \dfrac{2}{1-4x^2} dx$.

Let $u = 2x; \; du = 2;$

$\int \dfrac{2}{1-4x^2} dx = \int \dfrac{du}{1-u^2};$

We know that $D_x[\tanh^{-1} u] = \dfrac{D_x[u]}{1-u^2}$, so

$\int \dfrac{2}{1-4x^2} dx = \tanh^{-1}(2x) + c$

Congratulations! This is another advanced technique covered in Chapter 11.

Chapter 11: Advanced Integration Techniques

Pattern Recognition:

There are two final integration techniques to cover. The first one is rather easy.   Hopefully, you've discovered by now (otherwise, please take note) that much of the magic of mathematics is knowing a few rules and knowing how to apply them over and over to different problems.

Integration is no different. **Pattern recognition** is simply identifying the general rule from an integral table that applies to the problem at hand. We'll add a few more forms to our list of integrals before considering examples.

As you've moved through this book, you've been introduced to a number of integral solutions and integral solving methods. Many times, we've intuited the integral solution from the derivative since the integral and the derivative are inverse functions of each other.

Recall that we determined the following derivative solutions of the inverse circular trig functions in Chapter 9:

$$D_x[\sin^{-1} u] = \frac{D_x[u]}{\sqrt{1-u^2}} = -D_x[\cos^{-1} u]$$

$$D_x[\tan^{-1} u] = \frac{D_x[u]}{1+u^2} = -D_x[\cot^{-1} u]$$

$$D_x[\sec^{-1} u] = \frac{D_x[u]}{u\sqrt{u^2-1}} = -D_x[\csc^{-1} u]$$

By inspection (i.e., knowing how inverse functions work), we can easily write the following:

$$\int \frac{du}{\sqrt{1-u^2}} = \sin^{-1} u + c \qquad -\int \frac{du}{\sqrt{1-u^2}} = \cos^{-1} u + c$$

$$\int \frac{du}{1+u^2} = \tan^{-1} u + c \qquad -\int \frac{du}{1+u^2} = \cot^{-1} u + c$$

$$\int \frac{du}{u\sqrt{u^2-1}} = \sec^{-1} u + c \qquad -\int \frac{du}{u\sqrt{u^2-1}} = \csc^{-1} u + c$$

Similarly, as we learned from Chapter 10:

$$D_x[\sinh^{-1}u] = \frac{1}{\sqrt{u^2+1}}\,du \qquad D_x[\operatorname{csch}^{-1}u] = \frac{-1}{u\sqrt{1+u^2}}\,du$$

$$D_x[\cosh^{-1}u] = \frac{1}{\sqrt{u^2-1}}\,du \qquad D_x[\operatorname{sech}^{-1}u] = \frac{-1}{u\sqrt{1-u^2}}\,du$$

$$D_x[\tanh^{-1}u] = \frac{1}{1-u^2}\,du \qquad D_x[\coth^{-1}u] = \frac{1}{1-u^2}\,du$$

Therefore, it's easy to see that:

$$\int \frac{1}{\sqrt{u^2+1}}\,du = \sinh^{-1}u + c \qquad \int \frac{-1}{u\sqrt{1+u^2}}\,du = \operatorname{csch}^{-1}u + c$$

$$\int \frac{1}{\sqrt{u^2-1}}\,du = \cosh^{-1}u + c \qquad \int \frac{-1}{u\sqrt{1-u^2}}\,du = \operatorname{sech}^{-1}u + c$$

$$\int \frac{1}{1-u^2}\,du = \tanh^{-1}u + c \qquad \int \frac{1}{1-u^2}\,du = \coth^{-1}u + c$$

Example

Evaluate $\int \frac{x}{\sqrt{4x^4+1}}\,dx$;

Let $u = 2x^2;\ u^2 = 4x^4;\ du = 4x$

$$\int \frac{x}{\sqrt{4x^4+1}}\,dx = \frac{1}{4}\int \frac{4x}{\sqrt{4x^4+1}}\,dx = \frac{1}{4}\int \frac{du}{\sqrt{u^2+1}}$$

$$= \frac{1}{4}\sinh^{-1}u + c = \frac{1}{4}\sinh^{-1}2x^2 + c$$

Example

Evaluate $\int \frac{2x}{x^4+4x^2+5}\,dx$;

The denominator almost looks like $u^2 + 1$:
$$u^2 + 1 = x^4 + 4x^2 + 4 + 1$$
$$u = \sqrt{x^4 + 4x^2 + 4)} = \sqrt{(x^2 + 2)^2};$$

$$u = x^2 + 2;\ du = 2x;$$

$$\int \frac{2x}{x^4+4x^2+5}\,dx = \int \frac{2x\,dx}{(x^2+2)^2+1} = \int \frac{du}{u^2+1}$$

$$= \tan^{-1}u + c = \tan^{-1}(x^2 + 2) + c$$

Partial Fraction Decomposition:

The last integration technique we'll work on is **partial fraction decomposition**. You may recall this technique from Algebra. In short, we're going to find a way to break up fractions in the integrand to make a subset of easier integrals to evaluate.

Recall the following rule about summing fractions:

$$\frac{A}{B} + \frac{C}{D} = \frac{AD+BC}{BD};$$

All we're going to do is work backward. We'll factor the denominator and then figure out what the numerator has to be. Then we'll be left with a simpler problem to solve.

To decompose a fraction into two partial fractions, we follow these steps:

1. Given $\frac{f(x)}{g(x)}$ we find B and D such that

$$g(x) = B \cdot D.$$

2. Rework the equation to look like:

$$\frac{f(x)}{g(x)} = \frac{f(x)}{BD} = \frac{A}{B} + \frac{C}{D} = \frac{AD+BC}{BD};$$

3. Finally, solve for A and C using:

$$f(x) = AD + BC$$

This is best shown by examples:

Example

Evaluate $\int \frac{3x}{x^2+2x-8} dx$;

$$\frac{3x}{x^2+2x-8} = \frac{3x}{(x+4)(x-2)} = \frac{A}{x+4} + \frac{B}{x-2}$$

$$= \frac{A(x-2)+B(x+4)}{(x+4)(x-2)}$$

This can only be true if:

$$3x = A(x-2) + B(x+4)$$

$$3x = (A+B)x - 2A + 4B$$

Grouping like terms gives us two equations with two unknowns:

$$3 = A + B;$$
$$0 = -2A + 4B \rightarrow A = 2B$$

$$3 = 3B \rightarrow B = 1, A = 2$$

Now the integral looks like this:

$$\int \frac{3x}{x^2+2x-8} dx = \int \left[\frac{2}{x+4} + \frac{1}{x-2}\right] dx$$

The problem has been converted to two very simple integrals:

$$\int \frac{3x}{x^2+2x-8} dx = \int \frac{2}{x+4} dx + \int \frac{1}{x-2} dx$$

$$= 2\ln|x+4| + \ln|x-2| + c$$

$$= \ln|(x-2)(x+4)^2| + c$$

Partial fraction decomposition can be done with more than two factors using similar techniques:

Evaluate $\int \frac{2x^2+6}{2x^3-4x^2-6x} dx$;

Factor the denominator:

$$\frac{2x^2+6}{2x^3-4x^2-6x} = \frac{2x^2+6}{2x(x^2-2x-3)} = \frac{2x^2+6}{2x(x-3)(x+1)}$$

$$= \frac{A}{2x} + \frac{B}{x-3} + \frac{C}{x+1}$$

$$= \frac{A(x-3)(x+1)+B(2x)(x+1)+C(2x)(x-3)}{2x(x-3)(x+1)}$$

Set numerators equal:

$$2x^2 + 6 =$$
$$A(x^2 - 2x - 3) + B(2x^2 + 2x) + C(2x^2 - 6x)$$

Solve for likes:

$$2 = A + 2B + 2C;$$
$$0 = -2A + 2B - 6C;$$
$$6 = -3A \rightarrow A = -2;$$

$$4 = 2B + 2C \rightarrow B = 2 - C;$$
$$-4 = 2(2 - C) - 6C \rightarrow C = 1;$$
$$B = 2 - 1 = 1;$$

Solve the three simpler integrals:

$$\int \frac{2x^2+6}{2x^3-4x^2-6x} dx = \int \frac{-2dx}{2x} + \int \frac{dx}{x-3} + \int \frac{dx}{x+1}$$

$$= -\ln|x| + \ln|x - 3| + \ln|x + 1| + c$$

$$= \ln\left|\frac{(x-3)(x+1)}{x}\right| + c$$

You may have noticed that in the above examples, the order of the polynomial in the numerator is less than the order of the polynomial in the denominator. This is important to check when doing decomposition. Keep in mind that in resolving a system into partial fractions, you want to find a form that will give simpler expressions of the type $\int \frac{du}{u}$ so that you can easily invoke the natural log solution.

Final Notes:

Understanding the story of calculus is a very important component of mastering the subject – one that can sometimes be overlooked. However, practice is equally as important. Though an individual tool may be easy to understand and apply, there are many, many tools to consider when attacking a particular problem. The task of finding the right method may at times seem overwhelming. Just take a breath, and follow these steps:

1. **Determine if you can directly use an integration technique such as:**
   - **Pull out constants and distribution**
   - **Functions of functions**
   - **Integration by parts**
   - **Partial fraction decomposition**

2. **Look for familiar functions:**
   - **Logs and exponentials**
   - **Circular trig functions and their inverses**
   - **Hyperbolic trig functions and their inverses**

3. **Look for patterns in the integrand and use look up tables for forms like:**
   - $u^2 \pm a^2; a^2 \pm u^2$
   - $\sqrt{u^2 \pm a^2}; \sqrt{a^2 \pm u^2}$

I wish you great success!

Appendices

A: Course Summary
B: Problem Sets
C: Solutions to Problem Sets
D: Derivative Tables
E: Integral Tables

Appendix A: Course Summary

1   Lines can be written in **slope intercept form**: $y = mx + b$

      Functions must pass the vertical line test.

2   Limit is defined as: $limit = L = \lim\limits_{x \to a} f(x)$

3   **Instantaneous slope = tangent = derivative =** $f'(x) = \lim\limits_{h \to 0} \frac{f(x+h)-f(x)}{h}$

      Basic rules of **derivatives**:

Derivatives of Constants are 0	$[c]' = 0$
Power Rule	$[x^n]' = nx^{n-1}$
Pull Out Constants	$[cf(x)]' = cf'$
Distribution	$[f \pm g]' = f' \pm g'$
Product Rule	$[f \cdot g]' = fg' + gf'$
Quotient Rule	$\frac{f'}{g'} = \frac{gf'-fg'}{g^2}$
Chain Rule	$\frac{dy}{dx} = \frac{dy}{du} \cdot \frac{du}{dx} = f'(u)du$

4   **Max/Min:**

To find an inflection point: Set $y' = 0$ and solve for x.

To determine max or min: Evaluate y'' at x values.

                                          If $y''(xi) > 0 \rightarrow$ min;

                                          If $y''(xi) < 0 \rightarrow$ max.

5   **Anti-derivatives** undo derivatives and are the process to evaluate integrals.

6   The **integral** expresses area and can be:

Indefinite: $\int f(x)dx = F(x) + c$

Definite: $\int_a^b f(x)dx = F(x)\big|_a^b = F(b) - F(a)$

7   Basic rules of **integration**:

    Change Sign/Flip Boundaries $\int_a^b f = -\int_b^a f$

    Pull Out Constants $\int_a^b cf = c\int_a^b f$

    Distribution $\int_a^b[f \pm g] = \int_a^b f \pm \int_a^b g$

    Break Intervals (a < k < b) $\int_a^b f = \int_a^k f + \int_k^b f$

    Functions of Functions $\int_a^b f(u)du = F(u)$

    Integration by Parts $\int_a^b u\,dv = uv - \int_a^b v\,du$

8,9   **Inverse functions** undo functions.

$$\log_a(a^x) = x;\ \ln e^x = x;\ \sin^{-1}(\sin x) = x;\ \sinh^{-1}(\sinh x) = x$$

10   **Hyperbolic functions** were defined:

$$\sinh x = \frac{e^x - e^{-x}}{2} \qquad\qquad \cosh x = \frac{e^x + e^{-x}}{2}$$

Derivatives & Integrals were defined and derived for:
          (Look up tables provide quick reference.)

8 **Logarithms**	9 **Circular Trig**	10 **Hyperbolic Trig**
Exponentials	**Inverse Circ Trig**	**Inverse Hyp Trig**

11   Advanced techniques include:

    Pattern recognition – Use look-up tables –
        Found in Appendices D & E as well as various other sources.

    Partial Fractions – Reduce to simpler problems

$$\frac{f(x)}{g(x)} = \frac{f(x)}{BD} = \frac{A}{B} + \frac{C}{D} = \frac{AD + BC}{BD}$$

Chapter correlations are given in large grey font.

Appendix B: Problem Sets

Chapter 1

Find the midpoint, distance, and slope for the following points:

1.1: (2,-2); (4,2)   **1.2:** (5,8); (1,2)   **1.3:** (-1,2); (1,-4)

1.4-1.6: Write the slope-intercept form for the line indicated by the above points.

Determine whether the following lines are parallel, perpendicular, or neither with the line $y = 4x + 2$:

1.7: $y = 2x - 3$ **1.8:** $y - 2 = 4(x - 3)$

1.9: $4y + x = 8$ **1.10:** $2y = 4x + 8$

Which of the following are functions?

1.11: $y = 7x^2 + 3x + 2$ **1.12:** $y = x^2 + \sin 2x$

1.13: $y = \begin{cases} 2x & x > 0 \\ 2x - 2 & x < 0 \end{cases}$ **1.14:** $y = \begin{cases} x^2 - 2 & x > 0 \\ x^3 + 3 & x < 1 \end{cases}$

State the order of the polynomial:

1.15: $y = 15x^5 + 4x^3 - 3$

1.16: $y = 5x^2 - 3x^3 + 4x - 1$

1.17: $y = x^{15} - 4x^{13} + 7x^{10} - 10x^8 + 2x^2 - 15$

Evaluate $(f \circ g)(x)$ and $(g \circ f)(x)$:

1.18: $f(x) = x^2; \ g(x) = 2x + 8$

1.19: $f(x) = \cos x; \ g(x) = 5x^2 + 1$

1.20: $f(x) = x^2 + 2; \ g(x) = 4x + 3$

Chapter 2

Find the limits:

2.1: $\lim\limits_{x \to 2} A$

2.2: $\lim\limits_{x \to -1} A$

2.3: $\lim\limits_{x \to \infty} B$

2.4: $\lim\limits_{x \to -\infty} B$

2.5: $\lim\limits_{x \to 0^+} C$

2.6: $\lim\limits_{x \to 0^-} C$

2.7: $\lim\limits_{x \to -1^+} D$

2.8: $\lim\limits_{x \to -1^-} D$

2.9: $\lim\limits_{x \to 0^+} E$

2.10: $\lim\limits_{x \to 0^-} E$

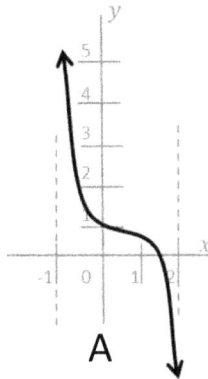

A

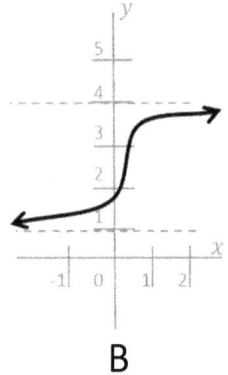

B

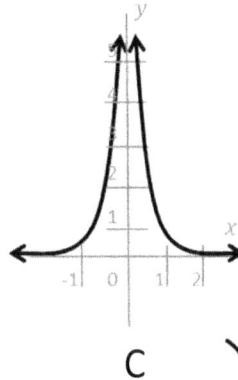

C

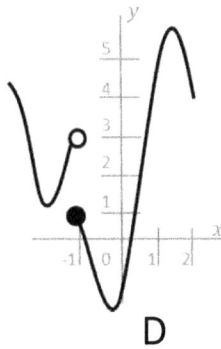

D

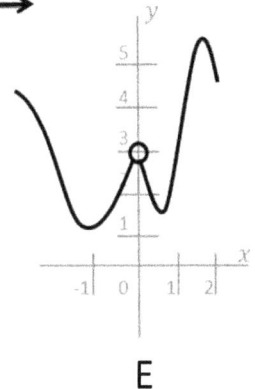

E

Find the instantaneous slope of the following functions at the x value given:

2.11: $y = 4x^3 - 3x;\ \ x = 0$

2.12: $y = x^2 + 5;\ \ x = 1$

2.13: $y = 4x^4 + x^2 + 5;\ \ x = 0$

2.14: $y = 4x^4;\ \ x = 1$

2.15: $y = x^2 - 5x + 4;\ \ x = 2$

2.16-2.20: Develop the tangent line equations for problems 2.11-2.15

Chapter 3

Find the derivatives of the following using the formal definition based on limits:

3.1: $y = 5x + 3$ **3.2:** $y = 4x^2 + 2$

3.3: $y = x^3 + 2x^2$ **3.4:** $y = x \sin x$

Find the derivatives of the following using rules for derivatives:

3.5: $y = 3x^2 + 4x + 5$ **3.6:** $y = 7x^3 + 14x^2 - 5x + 3$

3.7: $y = x^5 + \sqrt{x}$ **3.8:** $y = \frac{3}{x^2} - 2x$

3.9: $y = (3x + 5)^4 + (4x^2 + 2)^2$ **3.10:** $y = (4x + 2)(3x^2 + 5x + 3)$

3.11: $y = \frac{2x^3 + 3}{5x}$ **3.12:** $y = 2 \cos x \sin x$

3.13: $y = 5x + \sin 5x$ **3.14:** $y = \sin^2 5x$

3.15: $y = \cos x \tan x$ **3.16:** $y = (x^2 + 5)^2 \sin 5x$

3.17: $y = \frac{x \sin 2x}{x^2 + 1}$ **3.18:** $y = (2x^2 + 1)^2 \tan x^2$

Prove the following:

3.19: $D_x[\csc(x)] = -\csc(x) \cot(x)$

3.20: $D_x[\cot(x)] = -\csc^2(x)$

Chapter 4

Find the critical points of the following functions and determine whether they occur at a local maximum, minimum, or neither:

4.1: $y = 4x^2 + 2$ **4.2:** $y = x^3 + 2x^2 + 5$

4.3: $y = x^3 + 6x^2 + 3x + 14$ **4.4:** $y = 5x^2 + 3x + 2$

4.5: $y = x^4 + 2x^3 + x^2 + 3$ **4.6:** $y = \frac{x^3}{3} + 2x^2 - 5x + 3$

4.7: $y = \frac{x^3}{3} - 4x + 3$ **4.8:** $y = \frac{4x^3}{3} - \frac{3x^2}{2} - x + 2$

4.9: $y = x^2 - 10x + 5$ **4.10:** $y = \frac{x^3}{3} + \frac{x^2}{2} - 2x + 4$

4.11: $y = 2x^2 + 4x + 5$ **4.12:** $y = 3x^3 - 9x + 2$

4.13: $y = x^4 + 3x^3 + 2x^2$ **4.14:** $y = 9x^2 + 6x + 2$

4.15: $y = 4x - x^2$ **4.16:** $y = (2x + 1)^3$

4.17: $y = (4x^2 - 4)^2$ **4.18:** $y = 4x^3 - 12x$

4.19: $y = (x + 1)^3 - 3x$ **4.20:** $y = 8x^3 - 3x^2$

Chapter 5

Find $F(x)$:

5.1: $f(x) = x^2 + 3$

5.2: $f(x) = 4x^3 + 3x^2 + 5$

5.3: $f(x) = -8x^2 + 3$

5.4: $f(x) = \frac{-1}{\sqrt{x}}$

5.5: $f(x) = 2\cos 2x$

5.6: $f(x) = 10x^3 + 3x^2 + 5x + 2$

5.7: $f(x) = \sin(3x + 5)$

5.8: $f(x) = 9x^2 - \sin 4x$

5.9: $f(x) = (x^2 + 1)^2$

5.10: $f(x) = \frac{1}{x^2} - \frac{1}{x^3} + 3$

5.11: $f(x) = (x + 1)^3$

5.12: $f(x) = 21x^2 + 18x$

5.13: $f(x) = 3\cos(4x + 3)$

5.14: $f(x) = \left(x^2 - \frac{1}{x}\right)^2$

5.15: $f(x) = 4(x - 1)^3$

5.16: $f(x) = 6x^2 + 2x + 3$

Determine whether you can use the compound function rule, and if you can use the rule, find $F(x)$:

5.17: $f(x) = \sin(x^2 + 1)$

5.18: $f(x) = 4\cos(2x + 1)$

5.19: $f(x) = \tan \sqrt{x}$

5.20: $f(x) = (x^3 + 3)^2$

Chapter 6

Evaluate the integrals:

6.1: $\int 15x^2 + 2 \, dx$ **6.2:** $\int 4x^3 + 8x - 7 \, dx$

6.3: $\int 12\cos(3x) + 4 \, dx$ **6.4:** $\int 9x^5 - 4x^3 + 2x + 16 \, dx$

6.5: $\int_0^3 18x^2 + 10 \, dx$ **6.6:** $\int_0^1 9x^8 + 14x^6 - 20x^4 + 5 \, dx$

6.7: $\int_0^{\pi/4} 2\sin(2x) + 2 \, dx$ **6.8:** $\int_{-1}^1 3x^2 + 4x + 2 \, dx$

6.9: $\int_{-2}^0 5x^4 + 6x^2 + 9 \, dx$ **6.10:** $\int_{-3}^1 10x^4 - 6x^2 + 2 \, dx$

6.11: Evaluate $\int_0^2 x^3 + 3 \, dx$ by:

       a.) sum of 4 rectangles,   b.) sum of 8 rectangles, and   c.) direct integration.

Compare the difference of the integrals in first two columns to the value of the integral in the third column:

6.12: $\int_0^3 x^2 + 5x \, dx$ **6.13:** $\int_0^3 x + 4 \, dx$ **6.14:** $\int_0^3 x^2 + 4x - 4 \, dx$

6.15: $\int_0^2 5x^4 + 2x \, dx$ **6.16:** $\int_0^2 5x^4 + 3x^2 \, dx$ **6.17:** $\int_0^2 3x^2 - 2x \, dx$

6.18: $\int_1^2 6x^2 + 5 \, dx$ **6.19:** $\int_1^2 x^3 + 1 \, dx$ **6.20:** $\int_1^2 6x^2 - x^3 + 4 \, dx$

Chapter 7

Find the Area:

7.1:

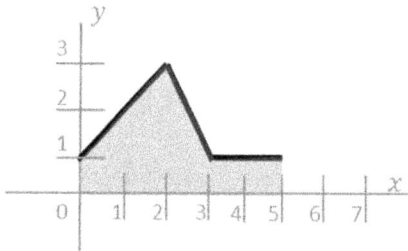

7.2:

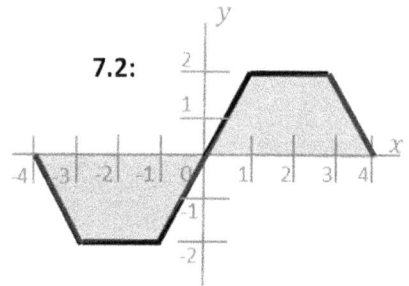

7.3: Between:
$y = 0; y = (x - 1)^2 - 1; x \in [0.4]$

7.4: Between:
$y = 1; y = 2 \sin x; x \in [0, \pi]$

7.5: Between:
$y = 8x - 2; y = 2x^3 - 2; x \in [-2,2]$

7.6: Between:
$y = 5x^4 - 3x^2; y = 4x^3 + 2x; x \in [-1,1]$

Evaluate the integrals:

7.7: $\int \sin(2x + 4)\, dx$

7.8: $\int x \cos 7x^2\, dx$

7.9: $\int \sin^2 3x \cos 3x\, dx$

7.10: $\int x^2 (4x^3 + 5)^4\, dx$

7.11: $\int \frac{x}{(x^2+5)^3}\, dx$

7.12: $\int (x^2 + 2) \sin(x^3 + 6x) dx$

7.13: $\int x^2 \sin(5x^3 + 3)\, dx$

7.14: $\int \csc^2 x \cos x\, dx$

7.15: $\int \frac{\sin \sqrt{2x}}{\sqrt{2x}}\, dx$

7.16: $\int (2x^2 - 1)(4x^3 - 3x^2 + 5)^4$

7.17: $\int 30x\,(x - 8)^4\, dx$

7.18: $\int 2x\,(x + 3)^3\, dx$

7.19: $\int (x + 2) \sin 2x\, dx$

7.20: $\int \frac{2x}{\sqrt{1-x}}\, dx$

Chapter 8

Evaluate the derivatives:

8.1: $D_x[5\ln(x^2+2)]$

8.2: $D_x[\ln(\sin x^3)]$

8.3: $D_x[3x^2\ln(x^3+4x)]$

8.4: $D_x[\log_2(5x^4+2x^2)]$

8.5: $D_x[e^{3x+9}]$

8.6: $D_x[e^{4x\sin 2x}]$

8.7: $D_x\left[5x^2e^{(x^2-6)^2}\right]$

8.8: $D_x[2^{(x^2+3)}]$

Evaluate the integrals:

8.9: $\int e^{3x+9}\,dx$

8.10: $\int \cos(2x)\,e^{4\sin(2x)}\,dx$

8.11: $\int xe^{2x}\,dx$

8.12: $\int x\,2^{(x^2-4)}\,dx$

8.13: $\int \frac{2x}{x^2+3}\,dx$

8.14: $\int \frac{x+2}{4x^2+16x+9}\,dx$

8.15: $\int \cot x\,dx$

8.16: $\int \frac{x+1}{x^2-x-2}\,dx$ Hint: Factor the denominator

8.17: $\int x\ln(4x^2-3)\,dx$

8.18: $\int 3x^2\ln(x^3+4)\,dx$

8.19: $\int \cos 2x\ln(\sin 2x)\,dx$

8.20: $\int x\log_4(x^2-4)\,dx$

Chapter 9

Evaluate the derivatives:

9.1: $D_x[\tan(\sqrt{x}+5)]$

9.2: $D_x[(x^2+2)\cot 4x^2]$

9.3: $D_x[\sin^3(3x^3)]$

9.4: $D_x[e^{x^2-2}\cos 4x]$

9.5: $D_x\left[\frac{\tan x^2}{\sec 5x}\right]$

9.6: $D_x[\cos^{-1}(1-x^2)]$

9.7: $D_x[\tan^{-1}(x+2)]$

9.8: $D_x\left[\sqrt{1-x^2}\sin^{-1}x\right]$

9.9: $D_x[e^{2x}\cot^{-1}(e^x)]$

9.10: $D_x\left[\frac{\tan^{-1}2x}{4x^2+1}\right]$

Evaluate the integrals:

9.11: $\int \sin x \cos^3 x\, dx$

9.12: $\int \sin(3x-9)\, dx$

9.13: $\int x\tan(x^2+5)\, dx$

9.14: $\int x\tan(x^2)\sec^4(x^2)\, dx$

9.15: $\int x\sec^2 x\, dx$

9.16: $\int x\csc^{-1}x^2\, dx$

9.17: $\int \cot^{-1}(4x+2)\, dx$

9.18: $\int x^2\cos^{-1}x^3\, dx$

9.19: $\int \cos x\, \tan^{-1}(\sin x)\, dx$

9.20: $\int e^x\csc^{-1}(e^x)\, dx$

Chapter 10

Evaluate the derivatives:

10.1: $D_x[\tanh(\sqrt{x}+5)]$

10.2: $D_x[(x^2+2)\coth 4x^2]$

10.3: $D_x[\sinh^3(3x^3)]$

10.4: $D_x[e^{x^2-2}\cosh 4x]$

10.5: $D_x\left[\frac{\tanh x^2}{\text{sech } 5x}\right]$

10.6: $D_x[\cosh^{-1}(1-x^2)]$

10.7: $D_x[\tanh^{-1}(x+2)]$

10.8: $D_x[\sqrt{1-x^2}\sinh^{-1}x]$

10.9: $D_x[e^{2x}\coth^{-1}(e^x)]$

10.10: $D_x\left[\frac{\tanh^{-1}2x}{4x^2+1}\right]$

Evaluate the integrals:

10.11: $\int \sinh x\cosh^3 x\,dx$

10.12: $\int \sinh(3x-9)\,dx$

10.13: $\int x\tanh(x^2+5)\,dx$

10.14: $\int x\tanh(x^2)\,\text{sech}^4(x^2)\,dx$

10.15: $\int x\,\text{sech}^2 x\,dx$

10.16: $\int x\,\text{csch}^{-1}x^2\,dx$

10.17: $\int \coth^{-1}(4x+2)\,dx$

10.18: $\int x^2\cosh^{-1}x^3\,dx$

10.19: $\int \cosh x\,\tanh^{-1}(\sinh x)\,dx$

10.20: $\int e^x\,\text{csch}^{-1}(e^x)\,dx$

Chapter 11

Evaluate the integrals:

11.1: $\int \frac{x}{\sqrt{1-x^4}} \, dx$

11.2: $\int \frac{\cos 3x}{1+\sin^2 3x} \, dx$

11.3: $\int \frac{1}{3x\sqrt{9x^2-1}} \, dx$

11.4: $\int \frac{x \cot x^2}{\sqrt{\sin^2 x^2 - 1}} \, dx$

11.5: $\int \frac{1}{e^x\sqrt{1-e^{-2x}}} \, dx$

11.6: $\int \frac{x}{\sqrt{x^4-1}} \, dx$

11.7: $\int \frac{\cos 3x}{1-\sin^2 3x} \, dx$

11.8: $\int \frac{1}{3x\sqrt{1-9x^2}} \, dx$

11.9: $\int \frac{x \cot x^2}{\sqrt{1+\sin^2 x^2}} \, dx$

11.10: $\int \frac{1}{e^x\sqrt{e^{-2x}-1}} \, dx$

11.11: $\int \frac{x-5}{x^2-x-2} \, dx$

11.12: $\int \frac{6x-10}{x^2-4x+3} \, dx$

11.13: $\int \frac{8x-2}{x^2-2x-8} \, dx$

11.14: $\int \frac{2x+1}{x^2+x-12} \, dx$

11.15: $\int \frac{-x}{x^2-x-20} \, dx$

11.16: $\int \frac{10x+5}{x^2-25} \, dx$

11.17: $\int \frac{-5(x+1)}{12x^2+2x-2} \, dx$

11.18: $\int \frac{6x^2+10x-30}{x^3+3x^2-10x} \, dx$

11.19: $\int \frac{3x^2+4x-1}{(x-1)(x^2+3x+2)} \, dx$

11.20: $\int \frac{4x^2-5x-2}{(x-1)(x^2-4x-4)} \, dx$

Appendix C: Solutions to Problem Sets

Chapter 1

	1.1: (2,-2); (4,2)	**1.2:** (5,8); (1,2)	**1.3:** (-1,2); (1,-4)
Midpoint	$\left(\frac{4+2}{2},\frac{-2+2}{2}\right)=(3,0)$	$\left(\frac{5+1}{2},\frac{8+2}{2}\right)=(3,5)$	$\left(\frac{-1+1}{2},\frac{2-4}{2}\right)=(0,-1)$
Distance	$\sqrt{(4-2)^2+(2+2)^2}$ $=\sqrt{20}$	$\sqrt{4^2+6^2}=\sqrt{52}$	$\sqrt{2^2+6^2}=\sqrt{40}$
Slope	$\frac{-2-2}{2-4}=\frac{-4}{-2}=2$	$\frac{8-2}{5-1}=\frac{6}{4}=1.5$	$\frac{-4-2}{1+1}=\frac{-6}{2}=-3$
	1.4: (2,-2); (4,2)	**1.5:** (5,8); (1,2)	**1.6:** (-1,2); (1,-4)
(0,b)	$y-b=m(x-0)$ $2-b=2(4)\rightarrow b=-6$	$2-b=1.5(1-0)$ $b=0.5$	$2-b=-3(-1-0)$ $b=-1$
Equation	$y=2x-6$	$y=1.5x+0.5$	$y=-3x-1$

To determine whether lines are parallel, perpendicular, or neither, compare the slopes. The slope of $y=4x+2$ is 4:

1.7: $y=2x-3$ $m=2$: Neither	**1.8:** $y-2=4(x-3)$ $m=4$: Parallel
1.9: $4y+x=8$ $y=-x/4+2$: $m=-1/4$: Perpendicular	**1.10:** $2y=4x+8$ $y=2x+4$: $m=2$: Neither

The easiest way to determine whether something is a function is to plot it and see if it passes the vertical line test:

1.11: $y = 7x^2 + 3x + 2$	**1.12:** $y = x^2 + \sin 2x$
Yes	Yes
1.13: $y = \begin{cases} 2x & x > 0 \\ 2x - 2 & x < 0 \end{cases}$	**1.14:** $y = \begin{cases} x^2 - 2 & x > 0 \\ x^3 + 3 & x < 1 \end{cases}$
Yes	No

The order of a polynomial is its highest exponent. By inspection:

1.15: $y = 15x^5 + 4x^3 - 3$	5
1.16: $y = 5x^2 - 3x^3 + 4x - 1$	3
1.17: $y = x^{15} - 4x^{13} + 7x^{10} - 10x^8 + 2x^2 - 15$	15

$(f \circ g)(x) = f(g(x))$ and $(g \circ f)(x) = g(f(x))$:

1.18: $f(x) = x^2$; $g(x) = 2x + 8$	$(f \circ g)(x)$ $= (2x + 8)^2$	$(g \circ f)(x)$ $= 2x^2 + 8$
1.19: $f(x) = \cos x$; $g(x) = 5x^2 + 1$	$(f \circ g)(x)$ $= \cos(5x^2 + 1)$	$(g \circ f)(x)$ $= 5\cos^2 x + 1$
1.20: $f(x) = x^2 + 2$; $g(x) = 4x + 3$	$(f \circ g)(x)$ $= (4x + 3)^2 + 2$	$(g \circ f)(x)$ $= 4x^2 + 11$

Chapter 2

Graphical limits can be determined by inspection:

2.1: $\lim\limits_{x \to 2} A = -\infty$

2.2: $\lim\limits_{x \to -1} A = \infty$

2.3: $\lim\limits_{x \to \infty} B = 4$

2.4: $\lim\limits_{x \to -\infty} B = 1$

2.5: $\lim\limits_{x \to 0^+} C = \infty$

2.6: $\lim\limits_{x \to 0^-} C = \infty$

2.7: $\lim\limits_{x \to -1^+} D = 1$

2.8: $\lim\limits_{x \to -1^-} D = 3$

2.9: $\lim\limits_{x \to 0^+} E = 3$

2.10: $\lim\limits_{x \to 0^-} E = 3$

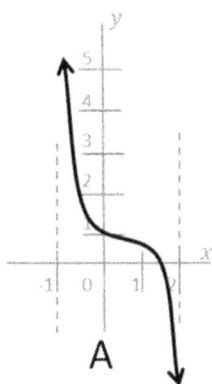

A

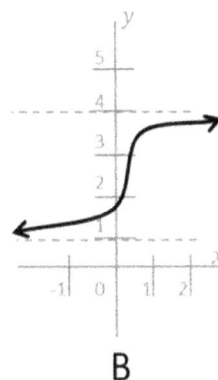

B

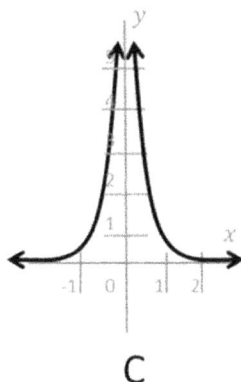

C

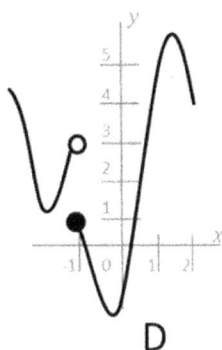

D

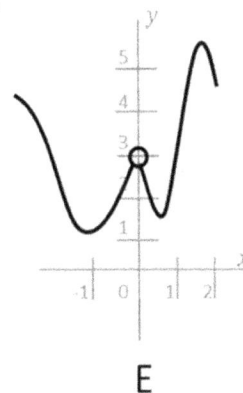

E

Use limits to find the instantaneous slope, and use the given x-value to find one point on the tangent line that can be used in the point-slope equation:

2.11: $y = 4x^3 - 3x$; $x = 0$ $\lim\limits_{x \to 0} \frac{4x^3 - 3x - 0 - 0}{x - 0} = \lim\limits_{x \to 0} 4x^2 - 3 = -3$	**2.16:** $y = 4x^3 - 3x$; $x = 0$ $(x, y) \to x = 0, y = 0$; $m = -3$ $y = -3x$
2.12: $y = x^2 + 5$; $x = 1$ $\lim\limits_{x \to 1} \frac{x^2 + 5 - 1 - 5}{x - 1} = \lim\limits_{x \to 1} \frac{x^2 - 1}{x - 1}$ $= \lim\limits_{x \to 1} \frac{(x-1)(x+1)}{x - 1} = \lim\limits_{x \to 1}(x + 1) = 2$	**2.17:** $y = x^2 + 5$; $x = 1$ $(x, y) \to x = 1, y = 6$; $m = 2$ $(y - 6) = 2(x - 1)$ $y = 2x + 4$
2.13: $y = 4x^4 + x^2 + 5$; $x = 0$ $\lim\limits_{x \to 0} \frac{4x^4 + x^2 + 5 - 0 - 0 - 5}{x - 0} = \lim\limits_{x \to 0} 4x^3 + x = 0$	**2.18:** $y = 4x^4 + x^2 + 5$; $x = 0$ $(x, y) \to x = 0, y = 5$; $m = 0$ $(y - 5) = 0(x - 0)$ $y = 5$
2.14: $y = 4x^4$; $x = 1$ $\lim\limits_{x \to 1} \frac{4x^4 - 4}{x - 1} = \lim\limits_{x \to 1} \frac{4(x^2 - 1)(x^2 + 1)}{x - 1}$ $= \lim\limits_{x \to 1} \frac{4(x-1)(x+1)(x^2 + 1)}{x - 1} = 4(2)(2) = 16$	**2.19:** $y = 4x^4$; $x = 1$ $(x, y) \to x = 1, y = 4$; $m = 16$ $(y - 4) = 16(x - 1)$ $y = 16x - 12$
2.15: $y = x^2 - 5x + 4$; $x = 2$ $\lim\limits_{x \to 2} \frac{x^2 - 5x + 4 - 4 + 10 - 4}{x - 2} = \lim\limits_{x \to 2} \frac{x^2 - 5x + 6}{x - 2}$ $= \lim\limits_{x \to 2} \frac{(x-2)(x-3)}{x - 2} = -1$	**2.20:** $y = x^2 - 5x + 4$; $x = 2$ $(x, y) \to x = 2, y = -2$; $m = -1$ $(y + 2) = -1(x - 2)$ $y = -x$

Chapter 3

Use the definition based on limits:

3.1: $y = 5x + 3$	**3.2:** $y = 4x^2 + 2$
$y' = \lim\limits_{h \to 0} \frac{5(x+h)+3-5x-3}{h}$ $= \lim\limits_{h \to 0} \frac{5x+5h-5x}{h} = 5$	$y' = \lim\limits_{h \to 0} \frac{4(x+h)^2+2-4x^2-2}{h}$ $= \lim\limits_{h \to 0} \frac{4x^2 + 8xh + 4h^2 - 4x^2}{h}$ $= \lim\limits_{h \to 0}(8x + h) = 8x$
3.3: $y = x^3 + 2x^2$ $y' = \lim\limits_{h \to 0} \frac{(x+h)^3+2(x+h)^2-x^3-2x^2}{h}$ $= \lim\limits_{h \to 0} \frac{x^3+3x^2h+3xh^2+h^3+2x^2+4xh+2h^2-x^3-2x^2}{h}$ $= \lim\limits_{h \to 0} \frac{3x^2h+3xh^2+h^3+4xh+2h^2}{h}$ $= 3x^2 + 4x$	**3.4:** $y = x \sin x$ $y' = \lim\limits_{h \to 0} \frac{(x+h)\sin(x+h)-x\sin x}{h}$ $= \lim\limits_{h \to 0} \frac{(x+h)(\sin x \cos h - \sin h \cos x) - x\sin x}{h}$ $= \lim\limits_{h \to 0}[\frac{x\sin x - x\sin x}{h} - x\cos x \frac{\sin h}{h}$ $+ \sin x - \sin h \cos x]$ $= \sin x - x\cos x$

3.5: $y = 3x^2 + 4x + 5$ $y' = 3(2)x + 4 = 6x + 4$	**3.6:** $y = 7x^3 + 14x^2 - 5x + 3$ $y' = 21x^2 + 28x - 5$
3.7: $y = x^5 + \sqrt{x}$ $y' = 5x^4 + \left(\frac{1}{2}\right)x^{-1/2} = 5x^4 + \frac{1}{2\sqrt{x}}$	**3.8:** $y = \frac{3}{x^2} - 2x$ $y' = 3(-2)x^{-3} - 2 = \frac{-6}{x^3} - 2$
3.9: $y = (3x + 5)^4 + (4x^2 + 2)^2$ $y' = 4(3x + 5)^3(3) + 2(4x^2 + 2)(8x)$ $y' = 12(3x + 5)^3 + 64x^3 + 32x$	**3.10:** $y = (4x + 2)(3x^2 + 5x + 3)$ $y' = (4x + 2)(6x + 5) + (3x^2 + 5x + 3)(4)$ $= 24x^2 + 32x + 10 + 12x^2 + 20x + 12$ $= 36x^2 + 52x + 22$
3.11: $y = \frac{2x^3 + 3}{5x}$ $y' = \frac{5x(6x^2) - (2x^3 + 3)(5)}{25x^2} = \frac{-10x^3 + 30x^2 - 15}{25x^2}$ $= \frac{-2x^3 + 6x^2 - 3}{5x^2}$	**3.12:** $y = 2\cos x \sin x$ $y' = 2[\cos x\,(\cos x) + \sin x\,(-\sin x)]$ $= 2[\cos^2 x - \sin^2 x]$

3.13: $y = 5x + \sin 5x$	**3.14:** $y = \sin^2 5x$
$y' = 5 + 5\cos 5x$	$y' = 2(\sin 5x)(5\cos 5x)$ $\quad = 10\sin 5x \cos 5x$
3.15: $y = \cos x \tan x$	**3.16:** $y = (x^2 + 5)^2 \sin 5x$
$y' = \cos x (\sec^2 x) - \tan x (\sin x)$	$y' = (x^2 + 5)^2 (5)\cos 5x$ $\quad + 2(x^2 + 5)(2x)\sin 5x$ $= 5(x^2 + 5)^2 \cos 5x$ $\quad + (4x^3 + 20x)\sin 5x$
3.17: $y = \dfrac{x \sin 2x}{x^2 + 1}$ $y' = \dfrac{(x^2+1)(x(2)\cos 2x + \sin 2x) + x \sin 2x\,(2x)}{(x^2+1)^2}$ $\quad = \dfrac{(x^2+1)(2x\cos 2x + \sin 2x) + 2x^2 \sin 2x}{(x^2+1)^2}$	**3.18:** $y = (2x^2 + 1)^2 \tan x^2$ $y' = (2x^2 + 1)^2 (\sec^2 x^2)(2x)$ $\quad + \tan x^2 (2)(2x^2 + 1)(4x)$ $\quad = (2x)(2x^2 + 1)^2 \sec^2 x^2$ $\quad + (8x)(2x^2 + 1)\tan x^2$
3.19: $D_x[\csc(x)] = -\csc(x)\cot(x)$	**3.20:** $D_x[\cot(x)] = -\csc^2(x)$
$y = \csc x = \dfrac{1}{\sin x}$ $y' = \dfrac{\sin x (0) - 1(\cos x)}{\sin^2 x} = \dfrac{-1}{\sin x}\left(\dfrac{\cos x}{\sin x}\right)$ $\quad = -\csc(x)\cot(x)$	$y = \cot x = \dfrac{\cos x}{\sin x}$ $y' = \dfrac{\sin x (-\sin x) - \cos x \cos x}{\sin^2 x}$ $\quad = \dfrac{-(\cos^2 x + \sin^2 x)}{\cos^2 x} = \dfrac{-1}{\cos^2 x} = -\csc^2(x)$

Chapter 4

4.1: $y = 4x^2 + 2$	**4.2:** $y = x^3 + 2x^2 + 5$
$$y' = 8x \equiv 0$$ $$x = 0 \therefore y = 2 \rightarrow (0, 2)$$ $$y'' = 8 > 0 \rightarrow Min$$	$$y' = 3x^2 + 4x \equiv 0$$ $$x(3x + 4) = 0 \rightarrow x = 0, -\,{}^4\!/_3$$ $$y'' = 6x + 4$$ $$(0, 5) \rightarrow y'' = 4 > 0 \rightarrow Min$$ $$(-{}^4\!/_3, 6.2) \rightarrow y'' = -4 < 0 \rightarrow Max$$
4.3: $y = x^3 + 6x^2 + 3x + 14$	**4.4:** $y = 5x^2 + 3x + 2$
$$y' = 3x^2 + 12x + 3 \equiv 0$$ $$x^2 + 4x + 1 = 0$$ $$x = \frac{-4 \pm \sqrt{16-4}}{2} = -0.26, -3.7$$ $$y'' = 6x + 12$$ $$(-0.26, 13.6) \rightarrow y'' = 10.4 > 0 \rightarrow Min$$ $$(-3.7, \ 34.4) \rightarrow y'' = -10.2 < 0 \rightarrow Max$$	$$y' = 10x + 3 \equiv 0$$ $$x = -0.3 \therefore y = 1.55 \rightarrow (-0.3, \ 1.55)$$ $$y'' = 10 > 0 \rightarrow Min$$
4.5: $y = x^4 + 2x^3 + x^2 + 3$	**4.6:** $y = \frac{x^3}{3} + 2x^2 - 5x + 3$
$$y' = 4x^3 + 6x^2 + 2x \equiv 0$$ $$2x(2x^2 + 3x + 1) = 0$$ $$2x(2x + 1)(x + 1) = 0$$ $$x = 0, -\,{}^1\!/_2, -1$$ $$y'' = 12x^2 + 12x + 2$$ $$(0, 3) \rightarrow y'' = 2 > 0 \rightarrow Min$$ $$(-{}^1\!/_2, 3.1) \rightarrow y'' = -1 < 0 \rightarrow Max$$ $$(-1, 3) \rightarrow y'' = 2 > 0 \rightarrow Min$$	$$y' = x^2 + 4x - 5 \equiv 0$$ $$(x + 5)(x - 1) = 0$$ $$x = 1, -5$$ $$y'' = 2x + 4$$ $$(1, 0.33) \rightarrow y'' = 6 > 0 \rightarrow Min$$ $$(-5, 36.3) \rightarrow y'' = -6 < 0 \rightarrow Max$$

4.7: $y = \frac{x^3}{3} - 4x + 3$	**4.8:** $y = \frac{4x^3}{3} - \frac{3x^2}{2} - x + 2$
$y' = x^2 - 4 \equiv 0$ $(x+2)(x-2) = 0 \rightarrow x = 2, -2$ $y'' = 2x$ $(2, -2.3) \rightarrow y'' = 4 > 0 \rightarrow Min$ $(-2, \ 8.3) \rightarrow y'' = -4 < 0 \rightarrow Max$	$y' = 4x^2 - 3x - 1 \equiv 0$ $(4x+1)(x-1) = 0 \rightarrow x = 1, -\frac{1}{4}$ $y'' = 8x - 3$ $(1, 0.83) \rightarrow y'' = 5 > 0 \rightarrow Min$ $(-0.25, \ 2.13) \rightarrow y'' = -5 < 0 \rightarrow Max$
4.9: $y = x^2 - 10x + 5$	**4.10:** $y = \frac{x^3}{3} + \frac{x^2}{2} - 2x + 4$
$y' = 2x - 10 \equiv 0$ $x = 5 \therefore y = -20 \rightarrow (5, \ -20)$ $y'' = 2 > 0 \rightarrow Min$	$y' = x^2 + x - 2 \equiv 0$ $(x+2)(x-1) = 0 \rightarrow x = 1, -2$ $y'' = 2x + 1$ $(1, 2.83) \rightarrow y'' = 3 > 0 \rightarrow Min$ $(-2, \ 7.33) \rightarrow y'' = -3 < 0 \rightarrow Max$
4.11: $y = 2x^2 + 4x + 5$	**4.12:** $y = 3x^3 - 9x + 2$
$y' = 4x + 4 \equiv 0$ $x = -1 \therefore y = 3 \rightarrow (-1, \ 3)$ $y'' = 4 > 0 \rightarrow Min$	$y' = 9x^2 - 9 \equiv 0$ $(x+1)(x-1) = 0 \rightarrow x = 1, -1$ $y'' = 18x$ $(1, -4) \rightarrow y'' = 18 > 0 \rightarrow Min$ $(-1, \ 8) \rightarrow y'' = -18 < 0 \rightarrow Max$
4.13: $y = x^4 + 3x^3 + 2x^2$	**4.14:** $y = 9x^2 + 6x + 2$
$y' = 4x^3 + 9x^2 + 4x \equiv 0$ $x(4x^2 + 9x + 2) = 0$ $x = 0, -1.64, -0.61$ $y'' = 12x^2 + 18x + 4$ $(0, 0) \rightarrow y'' = 2 > 0 \rightarrow Min$ $(-1.64, -0.62) \rightarrow y'' = 6.8 > 0 \rightarrow Min$ $(-0.61, 0.2) \rightarrow y'' = -2.5 < 0 \rightarrow Max$	$y' = 18x + 6 \equiv 0$ $x = -\frac{1}{3} \therefore y = 1 \rightarrow (-\frac{1}{3}, \ 1)$ $y'' = 18 > 0 \rightarrow Min$

4.15: $y = 4x - x^2$

$$y' = 4 - 2x \equiv 0$$
$$x = 2 \therefore y = 4 \rightarrow (2, \ 4)$$

$$y'' = -2 < 0 \rightarrow Max$$

4.16: $y = (2x + 1)^3$

$$y' = 3(2x + 1)^2 (2) \equiv 0$$
$$(2x + 1)^2 = 0 \rightarrow x = -\tfrac{1}{2}$$

$$y'' = (6)(2)(2x + 1)(2) = 48x + 24$$

$$(-\tfrac{1}{2}, 0) \rightarrow y'' = 0 \rightarrow Neither$$

The point found above
is simply an inflection
with no min or max

4.17: $y = (4x^2 - 4)^2$

$$y' = 2(4x^2 - 4)(8x) \equiv 0$$
$$x(x + 1)(x - 1) = 0 \rightarrow x = 0, 1, -1$$

$$y'' = D_x[64(x^3 - x)] = 192x^2 - 64$$

$$(0, 16) \rightarrow y'' = -64 < 0 \rightarrow Max$$
$$(1, \ 0) \rightarrow y'' = 128 > 0 \rightarrow Min$$
$$(-1, \ 0) \rightarrow y'' = 128 > 0 \rightarrow Min$$

4.18: $y = 4x^3 - 12x$

$$y' = 12x^2 - 12 \equiv 0$$
$$x = 1, -1$$

$$y'' = 24x$$

$$(1, -8) \rightarrow y'' = 24 > 0 \rightarrow Min$$
$$(-1, \ 8) \rightarrow y'' = -24 < 0 \rightarrow Max$$

4.19: $y = (x + 1)^3 - 3x$

$$y' = 3(x + 1)^2 - 3 \equiv 0$$
$$(x + 1)^2 = 1 \rightarrow x = 0, -2$$

$$y'' = 6(x + 1)$$

$$(0, 1) \rightarrow y'' = 6 > 0 \rightarrow Min$$
$$(-2, \ 2) \rightarrow y'' = -6 < 0 \rightarrow Max$$

4.20: $y = 8x^3 - 3x^2$

$$y' = 24x^2 - 6x \equiv 0$$
$$x(4x - 1) = 0 \rightarrow x = 0, \tfrac{1}{4}$$

$$y'' = 48x - 6$$

$$(0, 0) \rightarrow y'' = -6 < 0 \rightarrow Max$$
$$(\tfrac{1}{4}, \ -\tfrac{1}{16}) \rightarrow y'' = 6 > 0 \rightarrow Min$$

Chapter 5

5.1: $f(x) = x^2 + 3$ $$F(x) = \frac{x^3}{3} + 3x + c$$	**5.2:** $f(x) = 4x^3 + 3x^2 + 5$ $$F(x) = \frac{4x^4}{4} + \frac{3x^3}{3} + 5x + c$$ $$= x^4 + x^3 + 5x + c$$
5.3: $f(x) = -8x^2 + 3$ $$F(x) = \frac{-8x^3}{3} + 3x + c$$	**5.4:** $f(x) = \frac{-1}{\sqrt{x}}$ $$F(x) = -1(x)^{\frac{1}{2}} \div \left(\frac{1}{2}\right) + c$$ $$= -2\sqrt{x} + c$$
5.5: $f(x) = 2\cos 2x$ $$u = 2x; \quad du = 2$$ $$F(x) = \frac{F(u)}{du} = \frac{2\sin 2x}{2} = \sin 2x + c$$	**5.6:** $f(x) = 10x^3 + 3x^2 + 5x + 2$ $$F(x) = \frac{10x^4}{4} + \frac{3x^3}{3} + \frac{5x^2}{2} + 2x + c$$ $$= \frac{5}{2}(x^4 + x^2) + x^3 + 2x + c$$
5.7: $f(x) = \sin(3x + 5)$ $$u = 3x + 5; \quad du = 3$$ $$F(x) = \frac{F(u)}{du} = \frac{-\cos(3x+5)}{3} + c$$	**5.8:** $f(x) = 9x^2 - \sin 4x$ $$u = 4x; \quad du = 4$$ $$F(x) = 3x^3 + \frac{\cos 4x}{4} + c$$
5.9: $f(x) = (x^2 + 1)^2$ $$f(x) = x^4 + 2x^2 + 1$$ $$F(x) = \frac{x^5}{5} + \frac{2x^3}{3} + x + c$$	**5.10:** $f(x) = \frac{1}{x^2} - \frac{1}{x^3} + 3$ $$F(x) = \frac{x^{-1}}{-1} - \frac{x^{-2}}{-2} + 3x + c$$ $$= \frac{-1}{x} + \frac{1}{2x^2} + 3x + c$$

5.11: $f(x) = (x+1)^3$ $u = x+1; \quad du = 1$ $F(x) = \frac{F(u)}{du} = \frac{(x+1)^4}{4(1)} = \frac{(x+1)^4}{4} + c$	**5.12:** $f(x) = 21x^2 + 18x$ $F(x) = \frac{21x^3}{3} + \frac{18x^2}{2} + c$ $= 7x^3 + 9x^2 + c$
5.13: $f(x) = 3\cos(4x+3)$ $u = 4x+3; \quad du = 4$ $F(x) = \frac{F(u)}{du} = \frac{3\sin(4x+3)}{4} + c$	**5.14:** $f(x) = \left(x^2 - \frac{1}{x}\right)^2$ $f(x) = x^4 - 2x + x^{-2}$ $F(x) = \frac{x^5}{5} - x^2 - \frac{1}{x} + c$
5.15: $f(x) = 4(x-1)^3$ $u = x-1; \quad du = 1$ $F(x) = \frac{F(u)}{du} = \frac{4(x-1)^4}{4} = (x-1)^4 + c$	**5.16:** $f(x) = 6x^2 + 2x + 3$ $F(x) = \frac{6x^3}{3} + \frac{2x^2}{2} + 3x + c$ $= 2x^3 + x^2 + 3x + c$
5.17: $f(x) = \sin(x^2 + 1)$ $u = x^2 + 1; \quad du = 2x \neq constant$ You cannot use the rule in this case.	**5.18:** $f(x) = 4\cos(2x+1)$ $u = 2x+1; \quad du = 2$ $F(x) = \frac{4\sin(2x+1)}{2} = 2\sin(2x+1) + c$
5.19: $f(x) = \tan\sqrt{x}$ $u = \sqrt{x}; \quad du = \frac{1}{2\sqrt{x}} \neq constant$ You cannot use the rule in this case.	**5.20:** $f(x) = (x^3 + 3)^2$ $u = x^3 + 3; \quad du = 3x^2 \neq constant$ You have to multiply it out... $f(x) = x^6 + 6x^3 + 9$ $F(x) = \frac{x^7}{7} + \frac{3x^4}{2} + 9x + c$

Chapter 6

6.1: $\int 15x^2 + 2\ dx$ $= \frac{15x^3}{3} + 2x + c = 5x^3 + 2x + c$	**6.2:** $\int 4x^3 + 8x - 7\ dx$ $= \frac{4x^4}{4} + \frac{8x^2}{2} - 7x + c = x^4 + 4x^2 + c$
6.3: $\int 12\cos(3x) + 4\ dx$ $= 12\frac{\cos 3x}{3} + 4x + c = 4(\cos 3x + x) + c$	**6.4:** $\int 9x^5 - 4x^3 + 2x + 16\ dx$ $= \frac{9x^6}{6} - \frac{4x^4}{4} + \frac{2x^2}{2} + 16x + c$ $= \frac{3}{2}x^6 - x^4 + x^2 + 16x + c$
6.5: $\int_0^3 18x^2 + 10\ dx$ $= \frac{18x^3}{3} + 10x \Big|_0^3$ $= 6(3)^3 + 10(3) - [6(0)^3 + 10(0)] = 192$	**6.6:** $\int_0^1 9x^8 + 14x^6 - 20x^4 + 5\ dx$ $= \frac{(9x^9)}{9} + \frac{14x^7}{7} - \frac{20x^5}{5} + 5x \Big|_0^1$ $= 1^9 + 2(1)^7 - 4(1)^5 + 5(1) - 0 = 4$
6.7: $\int_0^{\pi/4} 2\sin(2x) + 2\ dx$ $= \frac{2\cos 2x}{2} + 2x \Big|_0^{\pi/4}$ $= \cos\left(\frac{\pi}{2}\right) + \frac{\pi}{2} - [\cos 0 + 2(0)] = 0.57$	**6.8:** $\int_{-1}^1 3x^2 + 4x + 2\ dx$ $= \frac{3x^3}{3} + \frac{4x^2}{2} + 2x \Big|_{-1}^1$ $= 1 + 2 + 2 - [-1 + 2 - 2] = 6$
6.9: $\int_{-2}^0 5x^4 + 6x^2 + 9\ dx$ $= \frac{5x^5}{5} + \frac{6x^3}{3} + 9x \Big|_{-2}^0$ $= 0 - [(-2)^5 + 2(-2)^3 + 9(-2)] = 66$	**6.10:** $\int_{-3}^1 10x^4 - 6x^2 + 2\ dx$ $= \frac{10x^5}{5} - \frac{6x^3}{3} + 2x \Big|_{-3}^1$ $= 2 - 2 + 2 - [486 + 54 - 6] = 440$

6.11: a.) Break the interval into 4 sections: $\Delta x = \frac{(2-0)}{4} = \frac{1}{2}$.

Next, find the x_i's that lie in the middle of each interval and calculate $f(x_i)$ for the height.

Finally, sum the areas of the rectangles:

$x_i =$	$^1/_4$	$^3/_4$	$^5/_4$	$^7/_4$
$x_i^3 + 3 =$	3.02	3.42	4.95	8.36

$Area =$ $\frac{1}{2}$[3.02+3.42+4.95+8.36] = 9.875

6.11: b.) Do the same thing except now $\Delta x = \frac{1}{4}$:

$x_i =$	$^1/_8$	$^3/_8$	$^5/_8$	$^7/_8$	$^9/_8$	$^{11}/_8$	$^{13}/_8$	$^{15}/_8$
$x_i^3 + 3 =$	3.00	3.05	3.24	3.67	4.42	5.60	7.29	9.59

$Area =$ $\frac{1}{4}$[3.00+3.05+3.24+3.67+4.42+5.60+7.29+9.59] = 9.967

6.11: c.) Direct integration gives:

$$\int_0^2 x^3 + 3\, dx = \frac{x^4}{4} + 3x\Big|_0^2 = \frac{16}{4} + 6 - 0 - 0 = 10$$

Notice that more rectangles give a closer approximation to the answer.

6.12: $\int_0^3 x^2 + 5x \, dx$	6.13: $\int_0^3 x + 4 \, dx$	6.14: $\int_0^3 x^2 + 4x - 4 \, dx$
$= \frac{x^3}{3} + \frac{5x^2}{2} \Big\|_0^3$	$= \frac{x^2}{2} + 4x \Big\|_0^3$	$= \frac{x^3}{3} + 2x^2 - 4x \Big\|_0^3$
$= \frac{27}{3} + \frac{45}{3} - \left[\frac{0}{3} + \frac{0}{3}\right]$	$= \frac{9}{2} + 12 - \left[\frac{0}{2} + 0\right]$	$= \frac{27}{3} + 2(9) - 12$
$= 31.5$	$= 16.5$	$- \left[\frac{0}{3} + 0 - 0\right]$
		$31.5 - 16.5 = 15$
6.15: $\int_0^2 5x^4 + 2x \, dx$	6.16: $\int_0^2 5x^4 + 3x^2 \, dx$	6.17: $\int_0^2 3x^2 - 2x \, dx$
$= x^5 + x^2 \Big\|_0^2$	$= x^5 + x^3 \Big\|_0^2$	$= x^3 - x^2 \Big\|_0^2 = 4$
$= 32 + 4 - 0 - 0$	$= 32 + 8 - 0 - 0$	$= 8 - 4$
$= 36$	$= 40$	$= 4$
		$40 - 36 = 4$
6.18: $\int_1^2 6x^2 + 5 \, dx$	6.19: $\int_1^2 x^3 + 1 \, dx$	6.20: $\int_1^2 6x^2 - x^3 + 4 \, dx$
$= 2x^3 + 5x \Big\|_1^2$	$= \frac{x^4}{4} + x \Big\|_1^2$	$= 2x^2 - \frac{x^4}{4} + 4x \Big\|_1^2 =$
$= 2(8) + 10 - [2 + 5]$	$= \frac{16}{4} + 2 - \left[\frac{1}{4} + 1\right]$	14.25
$= 19$	$= 4.75$	$= 8 - 4 + 8 - \left[2 - \frac{1}{4} + 4\right]$
		$= 14.25$
		$19 - 4.75 = 14.25$

Chapter 7

7.1:

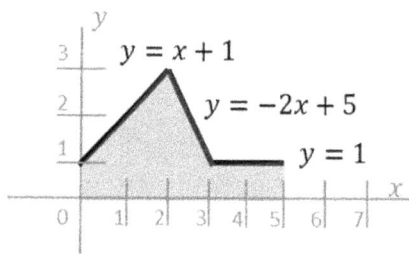

$$Area =$$

$$\int_0^2 x + 1 \, dx + \int_2^3 -2x + 5 \, dx + \int_3^5 dx$$

$$= \left[\frac{x^2}{2} + x\right]\Big|_0^2 + [-x^2 + 5x]\Big|_2^3 + x\Big|_3^5$$

$$= 2 + 2 - 9 + 15 + 4 - 10 + 5 - 3$$

$$= 9$$

7.2:

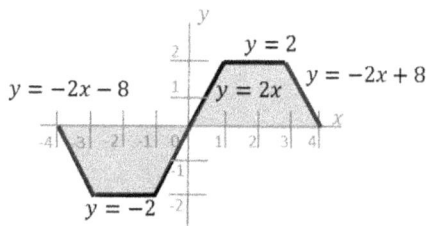

Remember, when the line is below the x-axis, the integral becomes $\int 0 - f(x) \, dx$

$$Area =$$

$$\int_{-4}^{-3} 2x + 8 \, dx + \int_{-3}^{-1} 2 \, dx + \int_{-1}^{0} -2x \, dx$$

$$+ \int_0^1 2x \, dx + \int_1^3 2 \, dx + \int_3^4 -2x + 8 \, dx$$

$$= [x^2 + 8x]\Big|_{-4}^{-3} + [2x]\Big|_{-3}^{-1} + [-x^2]\Big|_{-1}^{0}$$

$$+ [x^2]\Big|_0^1 + [2x]\Big|_1^3 + [-x^2 + 8x]\Big|_3^4$$

$$= 9 - 24 - 16 + 32 - 2 + 6 + 1$$
$$+ 1 + 6 - 2 - 16 + 32 + 9 - 24$$

$$= 12$$

7.3: Between:
$y = 0; y = (x - 1)^2 - 1; x \in [0.4]$

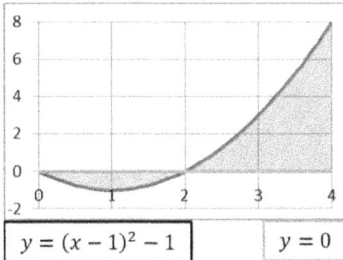

$y = (x - 1)^2 - 1$	$y = 0$

$(x - 1)^2 - 1 = 0 \rightarrow$ fn's cross at $x = 2$

$Area =$
$$\int_0^2 -(x - 1)^2 + 1 \, dx +$$
$$\int_2^4 (x - 1)^2 - 1 \, dx$$

$$= \frac{-(x-1)^3}{3} + x\Big|_0^2 + \frac{(x-1)^3}{3} - x\Big|_2^4$$

$$= \frac{-1}{3} + 2 - \frac{1}{3} + 9 - 4 - \frac{1}{3} + 2 = 8$$

7.4: Between:
$y = 1; y = 2 \sin x; x \in [0, \pi]$

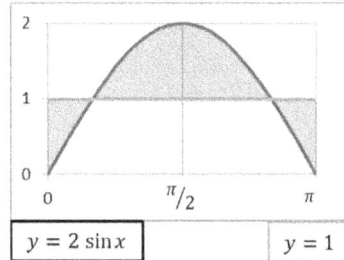

$y = 2 \sin x$	$y = 1$

$2 \sin x = 1 \rightarrow$ fn's cross at $x = \frac{\pi}{6}, \frac{5\pi}{6}$

$Area =$
$$\int_0^{\pi/6} 1 - 2 \sin x \, dx$$
$$+ \int_{\pi/6}^{5\pi/6} 2 \sin x - 1 \, dx + \int_{5\pi/6}^{\pi} 1 - 2 \sin x \, dx$$

$$= [x - 2 \cos x]\left[\Big|_0^{\pi/6} + \Big|_{5\pi/6}^{\pi}\right] + 2 \cos x - x \Big|_{\pi/6}^{5\pi/6}$$

$$= -1.2 + 2 + 5.14 - 1.8 + 1.2 - 1.8 = 3.54$$

7.5: Between:
$y = 8x - 2; y = 2x^3 - 2; x \in [-2,2]$

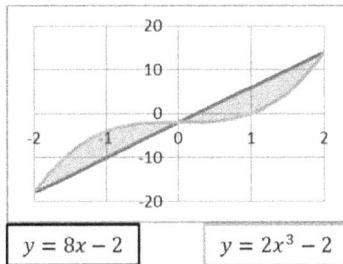

$y = 8x - 2$	$y = 2x^3 - 2$

$Area =$
$$\int_{-2}^0 2x^3 - 8x \, dx + \int_0^2 8x - 2x^3 \, dx$$

$$= \frac{x^4}{2} - 4x^2\Big|_{-2}^0 + 4x^2 - \frac{x^4}{2}\Big|_0^2$$

$$= -8 + 16 + 16 - 8 = 16$$

7.6: Between:
$y = 5x^4 - 3x^2; y = 4x^3 + 2x; x \in [-1,1]$

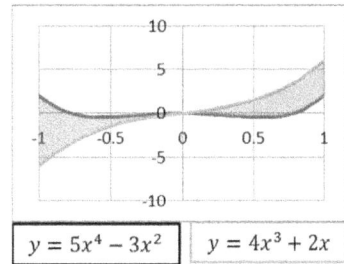

$y = 5x^4 - 3x^2$	$y = 4x^3 + 2x$

$Area =$
$$\int_{-1}^0 5x^4 - 3x^2 - 4x^3 - 2x \, dx$$
$$+ \int_0^1 4x^3 + 2x - 5x^4 + 3x^2 \, dx$$

$$= x^5 - x^3 - x^4 - x^2\Big|_{-1}^0$$

$$+ [-x^5 + x^3 + x^4 + x^2]\Big|_0^1 = 4$$

7.7: $\int \sin(2x + 4)\, dx$	**7.8:** $\int x \cos 7x^2\, dx$
$u = 2x + 4;\ \ du = 2\, dx;$	$u = 7x^2;\ \ du = 14x\, dx;$
$\frac{1}{2}\int 2\sin(2x+4)\, dx = \frac{1}{2}\int \sin u\, du$	$\frac{1}{14}\int 14 \cos 7x^2\, dx = \frac{1}{14}\int \cos u\, du$
$= \frac{-1}{2}\cos u + c = \frac{-1}{2}\cos(2x+4) + c$	$= \frac{1}{14}\sin u + c = \frac{1}{14}\sin 7x^2 + c$
7.9: $\int \sin^2 3x \cos 3x\, dx$	**7.10:** $\int x^2(4x^3 + 5)^4\, dx$
$u = \sin 3x;\ \ du = 3\cos 3x\, dx;$	$u = 4x^3 + 5;\ \ du = 12x^2\, dx;$
$\frac{1}{3}\int 3\sin^2 3x \cos 3x\, dx$	$\frac{1}{12}\int 12x^2(4x^3 + 5)^4 dx$
$= \frac{1}{3}\int u^2 du = \frac{1}{3}\frac{u^3}{3} = \frac{1}{9}\sin^3 3x + c$	$= \frac{1}{12}\int u^4 du = \frac{1}{12}\frac{u^5}{5} = \frac{(4x^3+5)^5}{60} + c$
7.11: $\int \frac{x}{(x^2+5)^3}\, dx$	**7.12:** $\int (x^2 + 2)\sin(x^3 + 6x)dx$
$u = x^2 + 5;\ \ du = 2x\, dx;$	$u = x^3 + 6x;\ \ du = 3x^2 + 6\, dx;$
	$\frac{1}{3}\int \sin u\, du = \frac{-1}{3}\cos u$
	$= \frac{-1}{3}\cos(x^3 + 6x) + c$
7.13: $\int x^2 \sin(5x^3 + 3)\, dx$	**7.14:** $\int \csc^2 x \cos x\, dx = \int \frac{\cos x}{\sin^2 x}\, dx$
$u = 5x^3 + 3;\ \ du = 15x^2\, dx;$	$u = \sin x;\ \ du = \cos x\, dx;$
$\frac{1}{15}\int \sin u\, du = \frac{-1}{15}\cos u$	$\int u^{-2} du = \frac{-1}{u} = \frac{-1}{\sin x} = -\csc x + c$
$= \frac{-1}{15}\cos(5x^3 + 3) + c$	

7.15: $\int \frac{\sin \sqrt{2x}}{\sqrt{2x}}\, dx$

$u = \sqrt{x}; \quad du = \frac{\sqrt{2}}{2\sqrt{x}}\, dx = \frac{1}{\sqrt{2x}}\, dx;$

$\int \sin u\, du = -\cos u = -\cos \sqrt{2x} + c$

7.16: $\int (2x^2 - 1)(4x^3 - 3x^2 + 5)^4$

$u = 4x^3 - 3x^2 + 5; \quad du = 12x^2 - 6x\, dx;$

$\frac{1}{6}\int u^4 du = \frac{1}{6}\frac{u^5}{5} = \frac{(4x^3 - 3x^2 + 5)^5}{30} + c$

7.17: $\int 30x\,(x - 8)^4\, dx$

Use: $\int u\,dv = uv - \int v\,du\,;$
$u = x; \quad du = dx;$
$dv = 30(x - 8)^4 dx = 30w^4 dw;$
$v = \frac{30w^5}{5} = 6w^5 = 6(x - 8)^5$

$\int u\,dv = 6x(x - 8)^5 - \int 6(x - 8)^5 dx$

$u = x - 8; \quad du = 1;$
$= 6x(x - 8)^5 - \int 6u^5 du$

$= 6x(x - 8)^5 - (x - 8)^6 + c$

7.18: $\int 2x\,(x + 3)^3\, dx$

Use: $\int u\,dv = uv - \int v\,du\,;$
$u = 2x; \quad du = 2dx;$
$dv = (x + 3)^3 dx = w^3 dw;$
$v = \frac{w^4}{4} = \frac{1}{4}(x + 3)^4$

$\int u\,dv = \frac{x}{2}(x + 3)^4 - \frac{1}{2}\int (x + 3)^4 du$

$= \frac{x}{2}(x + 3)^4 - \frac{1}{10}(x + 3)^5 + c$

7.19: $\int (x + 2)\sin 2x\, dx$

Use: $\int u\,dv = uv - \int v\,du\,;$
$u = x + 2; \quad du = dx;$
$dv = \sin 2x\, dx;$
$v = -\frac{\cos 2x}{2};$

$\int u\,dv = \frac{-(x+2)}{2}\cos 2x + \frac{1}{2}\int \cos 2x\, dx$

$u = 2x; du = 2dx;$

$= \frac{-(x+2)}{2}\cos 2x + \frac{1}{4}\sin 2x + c$

7.20: $\int \frac{2x}{\sqrt{1-x}}\, dx$

Use: $\int u\,dv = uv - \int v\,du\,;$
$u = x; \quad du = dx;$
$dv = 2(1 - x)^{-1/2}\, dx;$
$v = -\sqrt{1 - x};$

$\int u\,dv = -x\sqrt{1 - x} + \int \sqrt{1 - x}\,dx$

$u = 1 - x; du = -dx;$

$= -x\sqrt{1 - x} - \frac{2}{3}\sqrt{1 - x}^3 + c$

Chapter 8

8.1: $D_x[5\ln(x^2+2)]$	**8.2:** $D_x[\ln(\sin x^3)]$
$\quad = 5D_x[\ln u] = \dfrac{5}{u}du$	$\quad = D_x[\ln u] = \dfrac{1}{u}du$
$\quad = \dfrac{5(2x)}{x^2+2} = \dfrac{10x}{x^2+2}$	$\quad = \dfrac{3x^2}{\sin x^3}(\cos x^3) = 3x^2\cot x^3$
8.3: $D_x[3x^2\ln(x^3+4x)]$	**8.4:** $D_x[\log_2(5x^4+2x^2)]$
$\quad = 3x^2\left[\dfrac{3x^2+4}{x^3+4x}\right] + 6x\ln(x^3+4x)$	$\quad = D_x\log_2 u = \dfrac{du}{u\ln 2}$
	$\quad = \dfrac{20x^3+4x}{\ln 2\,[5x^4+2x^2]}$
8.5: $D_x[e^{3x+9}]$	**8.6:** $D_x[e^{4x\sin 2x}]$
$\quad D_x e^u = e^u du$	$\quad D_x e^u = e^u du$
$\quad = 3e^{3x+9}$	$\quad = (8x\cos 2x + 4\sin 2x)\,e^{4x\sin 2x}$
8.7: $D_x\left[5x^2 e^{(x^2-6)^2}\right]$	**8.8:** $D_x[2^{(x^2+3)}]$
$\quad = 5x^2\left[2(x^2-6)(2x)e^{(x^2-6)^2}\right]$	$\quad = D_x[2^u] = 2^u\ln 2\,du$
$\quad\quad +10xe^{(x^2-6)^2}$	$\quad = 2x\,(\ln 2)\,[2^{(x^2+3)}]$
$\quad = 10x\,e^{(x^2-6)^2}[2x^2(x^2-6)+1]$	

8.9: $\int e^{3x+9}\,dx$ $\quad u = 3x + 9; \quad du = 3\,dx;$ $\frac{1}{3}\int 3e^{3x+9}dx = \frac{1}{3}\int e^u du = \frac{e^u}{3} = \frac{e^{3x+9}}{3} + c$	**8.10:** $\int \cos(2x)\,e^{4\sin(2x)}\,dx$ $\quad u = 4\sin 2x; \quad du = 8\cos 2x\,dx;$ $\frac{1}{8}\int 8\cos(2x)\,e^{4\sin(2x)}\,dx = \frac{e^{4\sin 2x}}{8} + c$
8.11: $\int xe^{2x}\,dx$ $\quad u = x; \quad du = dx;$ $\quad dv = e^{2x}dx; \quad v = \frac{e^{2x}}{2};$ $\int u\,dv = \frac{xe^{2x}}{2} - \int \frac{e^{2x}}{2}\,dx = \frac{xe^{2x}}{2} - \frac{e^{2x}}{4} + c$	**8.12:** $\int x\,2^{(x^2-4)}\,dx$ $\quad u = x^2 - 4; \quad du = 2x\,dx;$ $\frac{1}{2}\int 2^u du = \frac{2^u}{2\ln 2} = \frac{2^{(x^2-4)}}{2\ln 2} + c$
8.13: $\int \frac{2x}{x^2+3}\,dx$ $\quad u = x^2 + 3; \quad du = 2x\,dx;$ $\int \frac{du}{u} = \ln|u| = \ln|x^2 + 3| + c$	**8.14:** $\int \frac{x+2}{4x^2+16x+9}\,dx$ $\quad u = 4x^2 + 16x + 9; \quad du = 8x + 16\,dx;$ $\frac{1}{8}\int \frac{du}{u} = \frac{\ln|u|}{8} = \frac{\ln|4x^2+16x+9|}{8} + c$
8.15: $\int \cot x\,dx = \int \frac{\cos x}{\sin x}\,dx$ $\quad u = \sin x; \quad du = \cos x\,dx;$ $\int \frac{du}{u} = \ln|u| = \ln|\sin x| + c$	**8.16:** $\int \frac{x+1}{x^2-x-2}\,dx = \int \frac{x+1}{(x+1)(x-2)}\,dx = \int \frac{dx}{x-2}$ $\quad u = x - 2; \quad du = dx;$ $\int \frac{du}{u} = \ln|u| = \ln|x - 2| + c$
8.17: $\int x\ln(4x^2 - 3)\,dx$ $\quad u = 4x^2 - 3; \quad du = 8x\,dx;$ $\quad \frac{1}{8}\int \ln u\,du = \frac{u}{8}[\ln u - 1]$ $= \frac{1}{8}(4x^2 - 3)(\ln|4x^2 - 3| - 1) + c$	**8.18:** $\int 3x^2\ln(x^3 + 4)\,dx$ $\quad u = x^3 + 4; \quad du = 3x^2\,dx;$ $\quad \int \ln u\,du = u[\ln u - 1]$ $= (x^3 + 4)(\ln|x^3 + 4| - 1) + c$
8.19: $\int \cos 2x\,\ln(\sin 2x)\,dx$ $\quad u = \sin 2x; \quad du = 2\cos 2x\,dx;$ $\frac{1}{2}\int \ln u\,du = \frac{\sin 2x}{2}(\ln|\sin 2x| - 1) + c$	**8.20:** $\int x\,\log_4(x^2 - 4)\,dx$ $\quad u = x^2 - 4; \quad du = 2x\,dx;$ $\frac{1}{2}\int \log_4 u\,du = \frac{x^2-4}{2}\left[\frac{\ln|x^2-4|}{\ln 4} - \ln 4\right] + c$

Chapter 9

9.1: $D_x[\tan(\sqrt{x}+5)]$ $\qquad D_x[\tan u] = \sec^2 u\, du\,;$ $\qquad = \frac{1}{2\sqrt{x}}\sec^2(\sqrt{x}+5)$	**9.2:** $D_x[(x^2+2)\cot 4x^2]$ $\qquad D_x[\cot u] = -\csc^2 u\, du\,;$ $\qquad = -8x(x^2+2)\csc^2 4x^2 + 2x\cot 4x^2$
9.3: $D_x[\sin^3(3x^3)]$ $\qquad D_x[\sin u] = \cos u\, du\,;$ $\qquad = (3\sin^2 3x^3)(9x^2)\cos 3x^3$ $\qquad = (27x^2)\cos 3x^3\sin^2 3x^3$	**9.4:** $D_x\left[e^{x^2-2}\cos 4x\right]$ $\qquad D_x[\cos u] = -\sin u\, du\,;$ $\qquad = e^{x^2-2}(-4)\sin 4x + \cos 4x\,(2x)e^{x^2-2}$ $\qquad = e^{x^2-2}(2x\cos 4x - 4\sin 4x)$
9.5: $D_x\left[\frac{\tan x^2}{\sec 5x}\right]$ $\qquad D_x[\sec] = \sec\tan du;\ D_x[\tan] = \sec^2 du\,;$ $\qquad = \frac{\sec 5x(\sec^2 x^2)(2x) - \tan x^2(5)\sec 5x\tan 5x}{\sec^2 5x}$ $\qquad = \frac{2x(\sec^2 x^2) - 5\tan 5x\tan x^2}{\sec 5x}$	**9.6:** $D_x[\cos^{-1}(1-x^2)]$ $\qquad D_x[\cos^{-1} u] = \frac{-du}{\sqrt{1-u^2}}\,;$ $\qquad = \frac{2x}{\sqrt{1-(1-x^2)^2}} = \frac{2x}{\sqrt{1-1+2x^2-x^4}} = \frac{2}{\sqrt{2-x^2}}$
9.7: $D_x[\tan^{-1}(x+2)]$ $\qquad D_x[\tan^{-1} u] = \frac{du}{1+u^2}\,;$ $\qquad = \frac{1}{1+(x+2)^2} = \frac{1}{x^2+4x+5}$	**9.8:** $D_x\left[\sqrt{1-x^2}\sin^{-1} x\right]$ $\qquad D_x[\sin^{-1} u] = \frac{du}{\sqrt{1-u^2}}\,;$ $\qquad = \frac{\sqrt{1-x^2}}{\sqrt{1-x^2}} + \frac{\sin^{-1} x}{\sqrt{1-x^2}}\left(\frac{-2x}{2}\right) = 1 - \frac{x\sin^{-1} x}{\sqrt{1-x^2}}$
9.9: $D_x[e^{2x}\cot^{-1}(e^x)]$ $\qquad D_x[\cot^{-1} u] = \frac{-du}{1+u^2}\,;$ $\qquad = \frac{-e^{2x}(e^x)}{1+e^{2x}} + 2e^{2x}\cot^{-1} e^x$	**9.10:** $D_x\left[\frac{\tan^{-1} 2x}{4x^2+1}\right]$ $\qquad = \frac{1}{(4x^2+1)^2}\left[\frac{(4x^2+1)(2)}{1+4x^2} - 8x\,\tan^{-1} 2x\right]$ $\qquad = \frac{1}{(4x^2+1)^2}[2 - 8x\,\tan^{-1} 2x]$

9.11: $\int \sin x \cos^3 x \, dx$

$\qquad u = \cos x; \quad du = -\sin x \, dx;$

$\qquad -\int u^3 du = -\frac{u^4}{4} = \frac{-1}{4}\cos^4 x + c$

9.12: $\int \sin(3x - 9) \, dx$

$\qquad u = 3x - 9; \quad du = 3 \, dx;$

$\qquad \frac{1}{3}\int \sin u \, du = -\frac{1}{3}\cos u = \frac{-\cos(3x-9)}{3} + c$

9.13: $\int x \tan(x^2 + 5) \, dx$

$\qquad u = x^2 + 5; \quad du = 2x \, dx;$

$\qquad \frac{1}{2}\int \tan u \, du = \frac{1}{2}\ln|\sec u|$

$\qquad = \frac{1}{2}\ln|\sec(x^2+5)| + c$

9.14: $\int x \tan(x^2) \sec^4(x^2) \, dx$

$\qquad u = \sec x^2; \quad du = 2x \sec x^2 \tan x^2 \, dx;$

$\qquad \frac{1}{2}\int u^3 du = \frac{u^4}{8} = \frac{1}{8}\sec^4 x^2 + c$

9.15: $\int x \sec^2 x \, dx$

$\qquad u = x; \quad du = dx;$

$\qquad dv = \sec^2 x \, dx; \quad v = \tan x;$

$\qquad \int u \, dv = x \tan x - \int \tan x \, dx$

$\qquad = x \tan x - \ln|\sec x| + c$

9.16: $\int x \csc^{-1} x^2 \, dx$

$\qquad u = x^2; \quad du = 2x \, dx;$

$\qquad \frac{1}{2}\int \csc^{-1} u \, du = \frac{u}{2}\csc^{-1} u + \frac{1}{2}\ln(u + \sqrt{u^2 - 1})$

$\qquad = \frac{1}{2}\left[x^2 \csc^{-1} x^2 + \ln(x^2 + \sqrt{x^4 - 1}\right] + c$

9.17: $\int \cot^{-1}(4x + 2) \, dx$

$\qquad u = 4x + 2; \quad du = 4 \, dx;$

$\qquad \frac{1}{4}\int \cot^{-1} u \, du = \frac{u}{4}\cot^{-1} u + \frac{1}{8}\ln|1 + u^2|$

$\qquad = \frac{4x+2}{4}\cot^{-1}(4x + 2) + \frac{\ln|1+(4x+2)^2|}{8} + c$

9.18: $\int x^2 \cos^{-1} x^3 \, dx$

$\qquad u = x^3; \quad du = 3x^2 dx;$

$\qquad \frac{1}{3}\int \cos^{-1} u \, du = \frac{1}{3}\left[u \cos^{-1} u - \sqrt{1 - u^2}\right]$

$\qquad = \frac{1}{3}\left[x^3 \cos^{-1} x^3 - \sqrt{1 - x^6}\right] + c$

9.19: $\int \cos x \tan^{-1}(\sin x) \, dx$

$\qquad u = \sin x; \quad du = \cos x \, dx;$

$\qquad \int \tan^{-1} u \, du = u \tan^{-1} u - \frac{1}{2}\ln|1 + u^2|$

$\qquad = \sin x \tan^{-1}(\sin x) - \frac{1}{2}\ln|1 + \sin^2 x| + c$

9.20: $\int e^x \csc^{-1}(e^x) \, dx$

$\qquad u = e^x; \quad du = e^x dx;$

$\qquad \int \csc^{-1} u \, du = u \csc^{-1} u + \ln(u + \sqrt{u^2 - 1})$

$\qquad = e^x \csc^{-1} e^x + \ln(e^x + \sqrt{e^{2x} - 1} + c$

Chapter 10

10.1: $D_x[\tanh(\sqrt{x}+5)]$ $D_x[\tanh u] = \text{sech}^2\, u\; du\,;$ $= \frac{1}{2\sqrt{x}}\text{sech}^2(\sqrt{x}+5)$	**10.2:** $D_x[(x^2+2)\coth 4x^2]$ $D_x[\coth u] = -\text{csch}^2 u\; du\,;$ $= -8x(x^2+2)\text{csch}^2 4x^2 + 2x\coth 4x^2$
10.3: $D_x[\sinh^3(3x^3)]$ $D_x[\sinh u] = \cosh u\; du\,;$ $= (3\sinh^2 3x^3)(9x^2)\cosh 3x^3$ $= (27x^2)\cosh 3x^3 \sinh^2 3x^3$	**10.4:** $D_x[e^{x^2-2}\cosh 4x]$ $D_x[\cosh u] = \sinh u\; du\,;$ $= e^{x^2-2}(4)\sinh 4x + \cosh 4x\,(2x)e^{x^2-2}$ $= e^{x^2-2}(2x\cosh 4x + 4\sinh 4x)$
10.5: $D_x\left[\frac{\tanh x^2}{\text{sech}\,5x}\right]$ $D_x[\text{sech}] = -\text{sech}\tanh du;\; D_x[\tan h] = \text{sech}^2\, du\,;$ $= \frac{\text{sech}\,5x(\text{sech}^2 x^2)(2x)+\tanh x^2(5)\,\text{sech}\,5x\tanh 5x}{\text{sech}^2 5x}$ $= \frac{2x(\text{sech}^2 x^2)+5\tanh 5x\tanh x^2}{\text{sech}\,5x}$	**10.6:** $D_x[\cosh^{-1}(1-x^2)]$ $D_x[\cosh^{-1} u] = \frac{du}{\sqrt{u^2-1}}\,;$ $= \frac{2x}{\sqrt{(1-x^2)^2-1}} = \frac{2x}{\sqrt{1-1-2x^2+x^4}} = \frac{2}{\sqrt{x^2-2}}$
10.7: $D_x[\tanh^{-1}(x+2)]$ $D_x[\tanh^{-1} u] = \frac{du}{1-u^2}\,;$ $= \frac{1}{1-(x+2)^2} = \frac{1}{-x^2-4x-3}$	**10.8:** $D_x[\sqrt{1-x^2}\,\sinh^{-1} x]$ $D_x[\sinh^{-1} u] = \frac{du}{\sqrt{1+u^2}}\,;$ $= \frac{\sqrt{1-x^2}}{\sqrt{1+x^2}} - \frac{x\sinh^{-1} x}{\sqrt{1-x^2}}$
10.9: $D_x[e^{2x}\coth^{-1}(e^x)]$ $D_x[\coth^{-1} u] = \frac{du}{1-u^2}\,;$ $= \frac{e^{2x}(e^x)}{1-e^{2x}} + 2e^{2x}\coth^{-1} e^x$	**10.10:** $D_x\left[\frac{\tanh^{-1} 2x}{4x^2+1}\right]$ $= \frac{1}{(4x^2+1)^2}\left[\frac{(4x^2+1)(2)}{1-4x^2} - 8x\tanh^{-1} 2x\right]$

10.11: $\int \sinh x \cosh^3 x \, dx$

$u = \cosh x; \quad du = \sinh x \, dx;$

$\int u^3 \, du = \frac{u^4}{4} = \frac{1}{4} \cosh^4 x + c$

10.12: $\int \sinh(3x - 9) \, dx$

$u = 3x - 9; \quad du = 3 \, dx;$

$\frac{1}{3} \int \sinh u \, du = \frac{1}{3} \cosh u = \frac{\cosh(3x-9)}{3} + c$

10.13: $\int x \tanh(x^2 + 5) \, dx$

$u = x^2 + 5; \quad du = 2x \, dx;$

$\frac{1}{2} \int \tanh u \, du = \frac{1}{2} \ln|\cosh u|$

$= \frac{1}{2} \ln|\cosh(x^2 + 5)| + c$

10.14: $\int x \tanh(x^2) \operatorname{sech}^4(x^2) \, dx$

$u = \operatorname{sech} x^2; \quad du = -2x \operatorname{sech} x^2 \tanh x^2 \, dx;$

$\frac{-1}{2} \int u^3 \, du = \frac{-u^4}{8} = \frac{-1}{8} \operatorname{sech}^4 x^2 + c$

10.15: $\int x \operatorname{sech}^2 x \, dx$

$u = x; \quad du = dx;$

$dv = \operatorname{sech}^2 x \, dx; \quad v = \tanh x;$

$\int u \, dv = x \tanh x - \int \tanh x \, dx$

$= x \tanh x - \ln|\cosh x| + c$

10.16: $\int x \operatorname{csch}^{-1} x^2 \, dx$

$u = x^2; \quad du = 2x \, dx;$

$\frac{1}{2} \int \operatorname{csch}^{-1} u \, du = \frac{u}{2} \operatorname{csch}^{-1} u + \frac{1}{2} \sinh^{-1} u$

$= \frac{1}{2} [x^2 \operatorname{csch}^{-1} x^2 + \sinh^{-1} x^2] + c$

10.17: $\int \coth^{-1}(4x + 2) \, dx$

$u = 4x + 2; \quad du = 4 \, dx;$

$\frac{1}{4} \int \coth^{-1} u \, du = \frac{u}{4} \coth^{-1} u + \frac{1}{8} \ln|u^2 - 1|$

$= \frac{4x+2}{4} \coth^{-1}(4x + 2) + \frac{\ln|(4x+2)^2 - 1|}{8} + c$

10.18: $\int x^2 \cosh^{-1} x^3 \, dx$

$u = x^3; \quad du = 3x^2 \, dx;$

$\frac{1}{3} \int \cosh^{-1} u \, du = \frac{1}{3} [u \cosh^{-1} u \mp \sqrt{u^2 - 1}]$

$= \frac{1}{3} [x^3 \cosh^{-1} x^3 \mp \sqrt{x^6 - 1}] + c$

10.19: $\int \cosh x \tanh^{-1}(\sinh x) \, dx$

$u = \sinh x; \quad du = \cosh x \, dx;$

$\int \tanh^{-1} u \, du = u \tanh^{-1} u + \frac{1}{2} \ln|1 - u^2|$

$= \sinh x \tanh^{-1}(\sinh x) + \frac{1}{2} \ln|1 - \sinh^2 x| + c$

10.20: $\int e^x \operatorname{csch}^{-1}(e^x) \, dx$

$u = e^x; \quad du = e^x \, dx;$

$\int \operatorname{csch}^{-1} u \, du = u \operatorname{csch}^{-1} u + \sinh^{-1} u$

$= e^x \operatorname{csch}^{-1} e^x + \sinh^{-1} e^x + c$

Chapter 11

11.1: $\int \frac{x}{\sqrt{1-x^4}}\,dx$ $u = x^2; \quad du = 2x\,dx$ $\frac{1}{2}\int \frac{du}{\sqrt{1-u^2}} = \frac{1}{2}\sin^{-1} u = \frac{1}{2}\sin^{-1} x^2 + c$	**11.2:** $\int \frac{\cos 3x}{1+\sin^2 3x}\,dx$ $u = \sin 3x; \quad du = 3\cos 3x\,dx;$ $\frac{1}{3}\int \frac{du}{1+u^2} = \frac{1}{3}\tan^{-1} u = \frac{1}{3}\tan^{-1}(\sin 3x) + c$
11.3: $\int \frac{1}{3x\sqrt{9x^2-1}}\,dx$ $u = 3x; \quad du = 3\,dx;$ $\frac{1}{3}\int \frac{du}{u\sqrt{u^2-1}} = \frac{1}{3}\sec^{-1} u = \frac{1}{3}\sec^{-1} 3x + c$	**11.4:** $\int \frac{x\cot x^2}{\sqrt{\sin^2 x^2-1}}\,dx = \int \frac{x\cos x^2}{\sin x^2\sqrt{\sin^2 x^2-1}}$ $u = \sin x^2; \quad du = 2x\cos x^2\,dx;$ $\frac{1}{2}\int \frac{du}{u\sqrt{u^2-1}} = \frac{1}{2}\sec^{-1}(\sin x^2) + c$
11.5: $\int \frac{1}{e^x\sqrt{1-e^{-2x}}}\,dx = \int \frac{e^{-x}dx}{\sqrt{1-e^{-2x}}}$ $u = e^{-x}; \quad du = -e^{-x}dx;$ $\int \frac{-du}{\sqrt{1-u^2}} = \cos^{-1} u = \cos^{-1} e^{-x} + c$	**11.6:** $\int \frac{x}{\sqrt{x^4-1}}\,dx$ $u = x^2; \quad du = 2x\,dx$ $\frac{1}{2}\int \frac{du}{\sqrt{u^2-1}} = \frac{1}{2}\cosh^{-1} u = \frac{1}{2}\cosh^{-1} x^2 + c$
11.7: $\int \frac{\cos 3x}{1-\sin^2 3x}\,dx$ $u = \sin 3x; \quad du = 3\cos 3x\,dx;$ $\frac{1}{3}\int \frac{du}{1-u^2} = \frac{1}{3}\tanh^{-1} u = \frac{1}{3}\tanh^{-1}(\sin 3x) + c$	**11.8:** $\int \frac{1}{3x\sqrt{1-9x^2}}\,dx$ $u = 3x; \quad du = 3\,dx;$ $\frac{1}{3}\int \frac{du}{u\sqrt{1-u^2}} = \frac{1}{3}\operatorname{sech}^{-1} u = \frac{1}{3}\operatorname{sech}^{-1} 3x + c$
11.9: $\int \frac{x\cot x^2}{\sqrt{1+\sin^2 x^2}}\,dx = \int \frac{x\cos x^2}{\sin x^2\sqrt{1+\sin^2 x^2}}\,dx$ $u = \sin x^2; \quad du = 2x\cos x^2\,dx;$ $\frac{1}{2}\int \frac{du}{u\sqrt{u^2+1}} = \frac{-1}{2}\operatorname{csch}^{-1}(\sin x^2) + c$	**11.10:** $\int \frac{1}{e^x\sqrt{e^{-2x}-1}}\,dx$ $u = e^{-x}; \quad du = -e^{-x}dx;$ $\int \frac{-du}{\sqrt{u^2-1}} = -\cosh^{-1} u = -\cosh^{-1} e^{-x} + c$

11.11: $\int \frac{x-5}{x^2-x-2}\,dx$

$$= \int \frac{x-5}{(x+1)(x-2)}\,dx \equiv \int \frac{A}{x+1} + \frac{B}{x-2}\,dx$$

$$= \int \frac{Ax-2A+Bx+B}{(x-1)(x-1)}\,dx$$

Solve for A,B:

$$x - 5 = Ax - 2A + Bx + B$$

$$1 = A + B \rightarrow A = 1 - B$$

$$-5 = -2A + B$$

$$-5 = -2 + 3B$$

$$B = -1$$

$$A = 2;$$

Integral becomes:

$$\int \frac{2}{x+1} + \frac{-1}{x-2}\,dx$$

$$u_1 = x + 1; \quad du_1 = dx;$$

$$u_2 = x - 2; \quad du_2 = dx;$$

$$\int \frac{x-5}{x^2-x-2}\,dx$$

$$= 2\ln|x+1| - \ln|x-2| + c$$

$$= \ln\left|\frac{(x+1)^2}{x-2}\right| + c$$

11.12: $\int \frac{6x-10}{x^2-4x+3}\,dx$

$$= \int \frac{6x-10}{(x-3)(x-1)}\,dx \equiv \int \frac{A}{x-3} + \frac{B}{x-1}\,dx$$

$$= \int \frac{Ax-A+Bx-3B}{(x-3)(x-1)}\,dx$$

Solve for A,B:

$$6x - 10 = Ax - A + Bx - 3B$$

$$6 = A + B \rightarrow A = 6 - B$$

$$-10 = -A - 3B$$

$$-10 = -6 - 2B$$

$$B = 2$$

$$A = 4;$$

Integral becomes:

$$\int \frac{4}{x-3} + \frac{2}{x-1}\,dx$$

$$u_1 = x - 3; \quad du_1 = dx;$$

$$u_2 = x - 1; \quad du_2 = dx;$$

$$\int \frac{6x-10}{x^2-4x+3}\,dx$$

$$= 4\ln|x-3| + 2\ln|x-1| + c$$

11.13: $\int \frac{8x-2}{x^2-2x-8}\,dx$

$= \int \frac{8x-2}{(x-4)(x+2)}\,dx \equiv \int \frac{A}{x-4} + \frac{B}{x+2}\,dx$

$= \int \frac{Ax+2A+Bx-4B}{(x-4)(x+2)}\,dx$

Solve for A,B:

$8x - 2 = Ax + 2A + Bx - 4B$

$8 = A + B \rightarrow A = 8 - B$

$-2 = 2A - 4B$

$-2 = 16 - 6B$

$B = 3$

$A = 5;$

Integral becomes:

$\int \frac{5}{x-4} + \frac{3}{x+2}\,dx$

$u_1 = x - 4;\ du_1 = dx;$

$u_2 = x + 2;\ du_2 = dx;$

$\int \frac{8x-2}{x^2-2x-8}\,dx$

$= 5\ln|x - 4| + 3\ln|x + 2| + c$

11.14: $\int \frac{2x+1}{x^2+x-12}\,dx$

$= \int \frac{2x+1}{(x-3)(x+4)}\,dx \equiv \int \frac{A}{x-3} + \frac{B}{x+4}\,dx$

$= \int \frac{Ax+4A+Bx-3B}{(x-3)(x+4)}\,dx$

Solve for A,B:

$2x + 1 = Ax + 4A + Bx - 3B$

$2 = A + B \rightarrow A = 2 - B$

$1 = 4A - 3B$

$1 = 8 - 7B$

$B = 1$

$A = 1;$

Integral becomes:

$\int \frac{1}{x-3} + \frac{1}{x+4}\,dx$

$u_1 = x - 3;\ du_1 = dx;$

$u_2 = x + 4;\ du_2 = dx;$

$\int \frac{2x+1}{x^2+x-12}\,dx$

$= \ln|x - 3| + \ln|x + 4| + c$

$= \ln|x^2 + x - 12| + c$

You get the same answer if you use:

$\int \frac{du}{u} = \ln|u| + c$

11.15: $\int \frac{-x}{x^2-x-20}\,dx$

$= \int \frac{-x}{(x-5)(x+4)}\,dx \equiv \int \frac{A}{x-5} + \frac{B}{x+4}\,dx$

$= \int \frac{Ax+4A+Bx-5B}{(x-5)(x+4)}\,dx$

Solve for A,B:

$-x = Ax + 4A + Bx - 5B$

$-1 = A + B \rightarrow A = B + 1$

$0 = 4A - 5B$

$0 = 4B + 4 - 5B$

$B = 4$

$A = 5;$

Integral becomes:

$\int \frac{5}{x-5} + \frac{4}{x+4}\,dx$

$u_1 = x - 5;\quad du_1 = dx;$

$u_2 = x + 4;\quad du_2 = dx;$

$\int \frac{-x}{x^2-x-20}\,dx$

$= 5\ln|x - 5| + 4\ln|x + 4| + c$

11.16: $\int \frac{10x+5}{x^2-25}\,dx$

$= \int \frac{10x+5}{(x-5)(x+5)}\,dx \equiv \int \frac{A}{x-5} + \frac{B}{x+5}\,dx$

$= \int \frac{Ax+5A+Bx-5B}{(x-5)(x+5)}\,dx$

Solve for A,B:

$10x + 5 = Ax + 5A + Bx - 5B$

$10 = 5(A + B) \rightarrow A = 2 - B$

$5 = 5A - 5B$

$5 = 10 - 10B$

$B = \frac{1}{2}$

$A = \frac{3}{2};$

Integral becomes:

$\int \frac{3}{2(x-5)} + \frac{B}{2(x+5)}\,dx$

$u_1 = x - 5;\quad du_1 = dx;$

$u_2 = x + 5;\quad du_2 = dx;$

$\int \frac{10x+5}{x^2-25}\,dx$

$= \frac{3}{2}\ln|x - 5| + \frac{1}{2}\ln|x + 5| + c$

$= \ln\left|\sqrt{x - 5}^3\right| + \ln|\sqrt{x + 5}| + c$

11.17: $\int \frac{-5(x+1)}{12x^2+2x-2} \, dx$

$= \int \frac{-5(x+1)}{(4x+2)(3x-1)} \, dx \equiv \int \frac{A}{4x+2} + \frac{B}{3x-1} \, dx$

$= \int \frac{3Ax-A+4Bx+2B}{(4x+2)(3x-1)} \, dx$

Solve for A,B:

$-5x - 5 = 3Ax - A + 4Bx + 2B$

$-5 = 3A + 4B$

$-5 = -A + 2B \rightarrow A = 2B + 5$

$-5 = 3(2B + 5) + 4B \rightarrow -20 = 10B$

$B = -2; \quad A = 1;$

Integral becomes:

$\int \frac{1}{4x+2} + \frac{-2}{3x-1} \, dx$

$u_1 = 4x + 2; \quad du_1 = 4 \, dx;$

$u_2 = 3x - 1; \quad du_2 = 3 \, dx;$

$\int \frac{-5(x+1)}{12x^2+2x-2} \, dx$

$= \frac{1}{4}\ln|4x + 2| - \frac{2}{3}\ln|3x - 1| + c$

11.18: $\int \frac{6x^2+10x-30}{x^3+3x^2-10x} \, dx$

$= \int \frac{6x^2+10x-30}{x(x+5)(x-2)} \, dx \equiv \int \frac{A}{x+5} + \frac{B}{x-2} + \frac{C}{x} \, dx$

$= \int \frac{Ax^2-2Ax+Bx^2+5Bx+Cx^2+3Cx-10C}{x(x+5)(x-2)} \, dx$

Solve for A,B,C:

$6x^2 + 10x - 30$

$= x^2(A + B + C) + x(-2A + 5B + 3C)$

$- 10C$

$6 = A + B + C \rightarrow A = 6 - B - C$

$10 = -2A + 5B + 3C$

$-30 = -10C \rightarrow C = 3$

$A = 3 - B$

$10 = -6 + 2B + 5B + 9 \rightarrow B = 1$

$A = 2;$

Integral becomes:

$\int \frac{2}{x+5} + \frac{1}{x-2} + \frac{3}{x} \, dx$

$u_1 = x + 5; \quad du_1 = dx;$

$u_2 = x - 2; \quad du_2 = dx;$

$\int \frac{6x^2+10x-30}{x^3+3x^2-10x} \, dx$

$= 2\ln|x + 5| + \ln|x - 2| + 3\ln|x| + c$

11.19: $\int \frac{3x^2+4x-1}{(x-1)(x^2+3x+2)} dx$

$$= \int \frac{3x^2+4x-1}{(x-1)(x+1)(x+2)} dx$$

$$\equiv \int \frac{A}{x+1} + \frac{B}{x-1} + \frac{C}{x+2} dx$$

$$= \int \frac{Ax^2+Ax-2A+Bx^2+3Bx+2B+Cx^2-C}{(x-1)(x+1)(x+2)} dx$$

Solve for A,B,C:

$$3x^2 + 4x - 1 = x^2(A + B + C) + x(A + 3B)$$
$$+ (-2A + 2B - C)$$

$$3 = A + B + C \rightarrow A + B = 3 - C$$
$$4 = A + 3B \rightarrow A = 4 - 3B$$
$$-1 = -2A + 2B - C$$

Add 1st and 2nd :

$$2 = -A + 3B$$
$$2 = -4 + 6B \rightarrow B = 1; A = 1$$
$$-1 = -2 + 2 - C \rightarrow C = 1;$$

Integral becomes:

$$\int \frac{1}{x+1} + \frac{1}{x-1} + \frac{1}{x+2} dx$$

$$\int \frac{3x^2+4x-1}{(x-1)(x^2+3x+2)} dx$$
$$= \ln|x + 1| + \ln|x - 1| + \ln|x + 2| + c$$

11.20: $\int \frac{4x^2-5x-2}{(x-1)(x^2-4x-4)} dx$

$$= \int \frac{4x^2-5x-2}{(x-1)(x-2)(x+2)} dx$$

$$\equiv \int \frac{A}{x-1} + \frac{B}{x-2} + \frac{C}{x+2} dx$$

$$= \int \frac{Ax^2-4A+Bx^2+Bx-2B+Cx^2-3Cx+2C}{(x-1)(x-2)(x+2)} dx$$

Solve for A,B,C:

$$4x^2 - 5x - 2 = x^2(A + B + C)$$
$$+x(B - 3C) - 4A - 2B + 2C$$

$$4 = A + B + C \rightarrow A + C = 4 - B$$
$$-5 = B - 3C \rightarrow B = 3C - 5$$
$$-2 = -4A - 2B + 2C$$
$$-2 = -4A - 4C + 10$$
$$3 = 4 - B \rightarrow B = 1$$
$$-6 = -3C \rightarrow C = 2$$
$$A = 4 - 2 - 1 = 1;$$

Integral becomes:

$$\int \frac{1}{x-1} + \frac{1}{x-2} + \frac{2}{x+2} dx$$

$$\int \frac{4x^2-5x-2}{(x-1)(x^2-4x-4)} dx$$
$$= \ln|x - 1| + \ln|x - 2| + 2\ln|x + 2| + c$$

Appendix D: Derivative Tables

Logarithms & Exponentials:

$$D_x[a^u] = a^u \ln a\, D_x[u] \qquad\qquad D_x[e^u] = e^u D_x[u]$$

$$D_x[\log_a u] = \frac{1}{u \ln a} D_x[u] \qquad\qquad D_x[\ln|u|] = \frac{1}{u} D_x[u]$$

Circular Trig Functions:

$$D_x[\sin(u)] = \cos(u)\, D_x[u] \qquad\qquad D_x[\cos(u)] = -\sin(u)\, D_x[u]$$

$$D_x[\tan(u)] = \sec^2(u)\, D_x[u] \qquad\qquad D_x[\sec(u)] = \sec(u)\tan(u)\, D_x[u]$$

$$D_x[\cot(u)] = -\csc^2(u)\, D_x[u \qquad\qquad D_x[\csc(u)] = -\csc(u)\cot(u)\, D_x[u]$$

Inverse Circular Trig Functions:

$$D_x[\sin^{-1} u] = \frac{D_x[u]}{\sqrt{1-u^2}} \qquad\qquad D_x[\cos^{-1} u] = \frac{-D_x[u]}{\sqrt{1-u^2}}$$

$$D_x[\tan^{-1} u] = \frac{D_x[u]}{1+u^2} \qquad\qquad D_x[\cot^{-1} u] = \frac{-D_x[u]}{1+u^2}$$

$$D_x[\sec^{-1} u] = \frac{D_x[u]}{u\sqrt{u^2-1}} \qquad\qquad D_x[\csc^{-1} u] = \frac{-D_x[u]}{u\sqrt{u^2-1}}$$

Hyperbolic Trig Functions:

$$D_x(\sinh u) = \cosh u\, D_x[u] \qquad\qquad D_x(\operatorname{csch} u) = -\operatorname{csch} u \coth u\, D_x[u]$$

$$D_x(\cosh u) = \sinh u\, D_x[u] \qquad\qquad D_x(\operatorname{sech} u) = -\operatorname{sech} u \tanh u\, D_x[u]$$

$$D_x(\tanh u) = \operatorname{sech}^2 u\, D_x[u] \qquad\qquad D_x(\coth u) = -\operatorname{csch}^2 u\, D_x[u]$$

Inverse Hyperbolic Trig Functions:

$$D_x[\sinh^{-1} u] = \frac{1}{\sqrt{u^2+1}} D_x[u] \qquad\qquad D_x[\operatorname{csch}^{-1} u] = \frac{-1}{u\sqrt{1+u^2}} D_x[u]$$

$$D_x[\cosh^{-1} u] = \frac{1}{\sqrt{u^2-1}} D_x[u] \qquad\qquad D_x[\operatorname{sech}^{-1} u] = \frac{-1}{u\sqrt{1-u^2}} D_x[u]$$

$$D_x[\tanh^{-1} u] = \frac{1}{1-u^2} D_x[u] \qquad\qquad D_x[\coth^{-1} u] = \frac{1}{1-u^2} D_x[u]$$

Appendix E: Integral Tables

Logarithms & Exponentials:

$$\int a^u du = \frac{a^u}{\ln a} + c$$

$$\int e^u du = e^u + c$$

$$\int a^u \ln a\, du = a^u + c$$

$$\int \frac{1}{u} du = \ln|u| + c$$

$$\int \log_a u\, du = u\left[\frac{\ln u}{\ln a} - \ln a\right] + c$$

$$\int \ln u\, du = u[\ln(u) - 1] + c$$

Circular Trig Functions:

$$\int \sin u\, du = -\cos u + c$$

$$\int \csc u\, du = \ln|\csc u - \cot u| + c$$

$$\int \cos u\, du = \sin u + c$$

$$\int \sec u\, du = \ln|\sec u + \tan u| + c$$

$$\int \tan u\, du = \ln|\sec u| + c$$

$$\int \cot u\, du = \ln|\sin u| + c$$

$$\int \sec^2 u\, du = \tan u + c$$

$$\int \sec u \tan u\, du = \sec u + c$$

$$\int \csc^2 u\, du = -\cot u + c$$

$$\int \csc u \cot u\, du = -\csc u + c$$

Inverse Circular Trig Functions:

$$\int \sin^{-1} u\, du = u \sin^{-1} u + \sqrt{1 - u^2} + c$$

$$\int \cos^{-1} u\, du = u \cos^{-1} u - \sqrt{1 - u^2} + c$$

$$\int \tan^{-1} u\, du = u \tan^{-1} u - \frac{1}{2}\ln|1 + u^2| + c$$

$$\int \csc^{-1} u\, du = u \csc^{-1} u \pm \ln(u + \sqrt{u^2 - 1}) + c \quad \begin{cases} +\to 0 < \csc^{-1} u < {}^\pi/_2 \\ -\to {}^{-\pi}/_2 < \csc^{-1} u < 0 \end{cases}$$

$$\int \sec^{-1} u\, du = u \sec^{-1} u \mp \ln(u + \sqrt{u^2 - 1}) + c \quad \begin{cases} -\to 0 < \sec^{-1} u < {}^\pi/_2 \\ +\to {}^\pi/_2 < \sec^{-1} u < \pi \end{cases}$$

$$\int \cot^{-1} u\, du = u \cot^{-1} u + \frac{1}{2}\ln|1 + u^2| + c$$

Hyperbolic Trig Functions:

$$\int \sinh u \, du = \cosh u + c$$

$$\int \cosh u \, du = \sinh u + c$$

$$\int \tanh u \, du = \ln|\cosh u| + c$$

$$\int \operatorname{sech}^2 u \, du = \tanh u + c$$

$$\int \operatorname{csch}^2 u \, du = -\coth u + c$$

$$\int \operatorname{csch} u \, du = \ln\left|\tanh\left(\frac{u}{2}\right)\right| + c$$

$$\int \operatorname{sech} u \, du = \tan^{-1}(\sinh u) + c$$

$$\int \coth u \, du = \ln|\sinh u| + c$$

$$\int \operatorname{sech} u \tanh u \, du = -\operatorname{sech} u + c$$

$$\int \operatorname{csch} u \coth u \, du = \operatorname{csch} u + c$$

Inverse Hyperbolic Trig Functions:

$$\int \sinh^{-1} u \, du = u \sinh^{-1} u - \sqrt{u^2 + 1} + c$$

$$\int \cosh^{-1} u \, du = u \cosh^{-1} u \mp \sqrt{u^2 - 1} + c \qquad \begin{cases} - \to \cosh^{-1} u > 0 \\ + \to \cosh^{-1} u < 0 \end{cases}$$

$$\int \tanh^{-1} u \, du = u \tanh^{-1} u + \frac{1}{2}\ln(1 - u^2) + c$$

$$\int \operatorname{csch}^{-1} u \, du = u \operatorname{csch}^{-1} u + \sinh^{-1} u + c$$

$$\int \operatorname{sech}^{-1} u \, du = u \operatorname{sech}^{-1} u + \sin^{-1} u + c$$

$$\int \coth^{-1} u \, du = u \coth^{-1} u + \frac{1}{2}\ln(u^2 - 1) + c$$

Forms with $u^2 + 1; u^2 - 1; 1 - u^2$:

$$\int \frac{du}{\sqrt{1-u^2}} = \sin^{-1} u + c$$

$$\int \frac{du}{1+u^2} = \tan^{-1} u + c$$

$$\int \frac{du}{u\sqrt{u^2-1}} = \sec^{-1} u + c$$

$$\int \frac{1}{\sqrt{u^2+1}} du = \sinh^{-1} u + c$$

$$\int \frac{1}{\sqrt{u^2-1}} du = \cosh^{-1} u + c$$

$$\int \frac{1}{1-u^2} du = \tanh^{-1} u + c$$

$$-\int \frac{du}{\sqrt{1-u^2}} = \cos^{-1} u + c$$

$$-\int \frac{du}{1+u^2} = \cot^{-1} u + c$$

$$-\int \frac{du}{u\sqrt{u^2-1}} = \csc^{-1} u + c$$

$$\int \frac{-1}{u\sqrt{1+u^2}} du = \operatorname{csch}^{-1} u + c$$

$$\int \frac{-1}{u\sqrt{1-u^2}} du = \operatorname{sech}^{-1} u + c$$

$$\int \frac{1}{1-u^2} du = \coth^{-1} u + c$$

Forms with $a^2 + u^2$; $a^2 - u^2$; $u^2 - a^2$:

$$\int \frac{du}{a^2+u^2} = \frac{1}{a}\tan^{-1}\frac{u}{a} + c \qquad\qquad \int \frac{u\,du}{a^2+u^2} = \frac{1}{2}\ln|a^2 + u^2| + c$$

$$\int \frac{du}{a^2-u^2} = \frac{1}{2a}\ln\left|\frac{u+a}{u-a}\right| + c \qquad\qquad \int \frac{u\,du}{a^2-u^2} = \frac{-1}{2}\ln|a^2 - u^2| + c$$

$$\int \frac{du}{u^2-a^2} = \frac{1}{2a}\ln\left|\frac{u-a}{u+a}\right| + c \qquad\qquad \int \frac{u\,du}{u^2-a^2} = \frac{1}{2}\ln|u^2 - a^2| + c$$

Forms with $\sqrt{a^2 + u^2}$:

$$\int \sqrt{a^2 + u^2}\,du = \frac{u}{2}\sqrt{a^2 + u^2} + \frac{a^2}{2}\ln\left|u + \sqrt{a^2 + u^2}\right| + c$$

$$\int \frac{\sqrt{a^2+u^2}}{u}\,du = \sqrt{a^2 + u^2} - a\ln\left|\frac{a+\sqrt{a^2+u^2}}{u}\right| + c$$

$$\int \frac{\sqrt{a^2+u^2}}{u^2}\,du = \frac{-\sqrt{a^2+u^2}}{u} + \ln\left|u + \sqrt{a^2 + u^2}\right| + c$$

$$\int \frac{du}{\sqrt{a^2+u^2}} = \ln\left|u + \sqrt{a^2 + u^2}\right| + c$$

$$\int \frac{du}{u\sqrt{a^2+u^2}} = \frac{-1}{a}\ln\left|\frac{a+\sqrt{a^2+u^2}}{u}\right| + c$$

$$\int \frac{du}{u^2\sqrt{a^2+u^2}} = \frac{-\sqrt{a^2+u^2}}{a^2u} + c$$

Forms with $\sqrt{a^2 - u^2}$:

$$\int \sqrt{a^2 - u^2}\, du = \frac{u}{2}\sqrt{a^2 - u^2} + \frac{a^2}{2}\sin^{-1}\frac{u}{a} + c$$

$$\int \frac{\sqrt{a^2 - u^2}}{u}\, du = \sqrt{a^2 - u^2} - a\ln\left|\frac{a + \sqrt{a^2 - u^2}}{u}\right| + c$$

$$\int \frac{\sqrt{a^2 - u^2}}{u^2}\, du = \frac{-\sqrt{a^2 - u^2}}{u} - \sin^{-1}\frac{u}{a} + c$$

$$\int \frac{du}{\sqrt{a^2 - u^2}} = \sin^{-1}\frac{u}{a} + c$$

$$\int \frac{du}{u\sqrt{a^2 - u^2}} = \frac{-1}{a}\ln\left|\frac{a + \sqrt{a^2 - u^2}}{u}\right| + c$$

$$\int \frac{du}{u^2\sqrt{a^2 - u^2}} = \frac{-\sqrt{a^2 - u^2}}{a^2 u} + c$$

Forms with $\sqrt{u^2 - a^2}$:

$$\int \sqrt{u^2 - a^2}\, du = \frac{u}{2}\sqrt{u^2 - a^2} - \frac{a^2}{2}\ln\left|u + \sqrt{u^2 - a^2}\right| + c$$

$$\int \frac{\sqrt{u^2 - a^2}}{u}\, du = \sqrt{u^2 - a^2} - a\cos^{-1}\frac{a}{u} + c$$

$$\int \frac{\sqrt{u^2 - a^2}}{u^2}\, du = \frac{-\sqrt{u^2 - a^2}}{u} + \ln\left|u + \sqrt{u^2 - a^2}\right| + c$$

$$\int \frac{du}{\sqrt{u^2 - a^2}} = \ln\left|u + \sqrt{u^2 - a^2}\right| + c$$

$$\int \frac{du}{u\sqrt{u^2 - a^2}} = \frac{-1}{a}\sec^{-1}\frac{u}{a} + c$$

$$\int \frac{du}{u^2\sqrt{u^2 - a^2}} = \frac{\sqrt{u^2 - a^2}}{a^2 u} + c$$

Index